T0137894

Signals and Communication Technology

Volume 218

For further volumes:
http://www.springer.com/series/4748

Mihai Dimian · Petru Andrei

Noise-Driven Phenomena in Hysteretic Systems

 Springer

Mihai Dimian
Department of Electrical and Computer
   Engineering
Howard University and Stefan cel Mare
   University
Washington, DC
USA

and

Stefan cel Mare University
Suceava
Romania

Petru Andrei
Department of Electrical and Computer
   Engineering
Florida State University and Florida
   A&M University
Tallahasse, FL
USA

ISBN 978-1-4939-5246-5   ISBN 978-1-4614-1374-5   (eBook)
DOI 10.1007/978-1-4614-1374-5
Springer New York Heidelberg Dordrecht London

Printed on acid-free paper

Springer is part of Springer Science+Business Media (www.springer.com)

Preface

Hysteresis and noise are ubiquitous phenomena in science and engineering playing an increasingly important role in research and technological developments. Mathematical modeling and simulations have flourished in both areas laying down a multidisciplinary framework for their analysis. While the quasistatic study of hysteresis has reached a certain degree of maturity reflected in the extensive monographs published over the last 30 years, the stochastic analysis of hysteretic systems is currently under major developments. Numerous papers dealing with various manifestations of noise in hysteretic phenomena have been recently published but their systematic analysis has been rather limited. This book is aimed at providing a general approach to nonlinear systems with hysteresis driven by noisy inputs, which leads to a unitary framework for the analysis of various stochastic aspects of hysteresis. An open access simulation platform with the models implementation can be used by the readers for learning and researching many topics discussed here.

This monograph is written for applied scientists and engineers who are interested in noise or in hysteretic systems. Familiarity with noise and hysteresis phenomena is recommended but not necessary for the understanding of the book. Many references are provided at the end of each chapter, which should help the reader understand the required mathematical background or the practical applications when necessary. In addition, the book should be useful to researchers or graduate students willing to implement noise and hysteresis models or to study the effects of hysteresis and noise in various applications originating from the classical areas of magnetism and plasticity as well as from superconductivity, photonics, computer and communication networks, economics, hydrology, biology and neuroscience.

Due to the wide spectrum of areas in which hysteresis is observed and to the fact that the origins of hysteresis are often multiple and unclear, there exist a large number of mathematical models for hysteresis in the literature. The first chapter offers an overview of the hysteresis models that are being used throughout the book. The chapter starts with a general classification of hysteresis models, parameter identification, and inverse modeling techniques. The rectangular hysteresis operator is first introduced being the building block of more complex hysteresis models. Then, the chapter focuses on summarizing the main equations, properties, and characteristics of the Preisach, energetic, Jiles-Atherton, Coleman-Hodgdon, and Bouc-Wen models. Particular attention is given to the analytical description of the general properties of hysteresis curves such as differential susceptibilities, remanence, coercivity, saturation, anhysteretic curve, energy loss, stability, accommodation, and limit cycle. The last two sections of the chapter are dedicated to dynamic and vector models of hysteresis. Two general techniques for modeling of rate-dependent hysteresis are presented, one based on the feedback (effective field) theory and the other on the relaxation time approximation. At the end, the chapter provides a general formalism that can be used to generalize scalar models of hysteresis in order to describe two-dimensional and three-dimensional vector systems.

The presentation in the first chapter does not make any reference to noise or other stochastic processes and could be also used by the readers as an introduction to the area of deterministic hysteretic modeling. This chapter may serve as reference for people working on finite element techniques in which the system properties are described by a hysteresis model. After reading Chap. 1, one might also be interested in reading the Appendix, which provides a description of these models implementation in HysterSoft©—an open access simulation platform used throughout the book.

Chapter 2 describes common noise models employed in various areas of research. In magnetic hysteresis, Gaussian white noise is mostly used as noise model and is mathematically defined by independent and identically distributed random variables following a Gaussian distribution. Once the interest in hysteresis extended beyond the traditional area of magnetism, various noise models appeared naturally in the hysteresis analysis. For example, pink noise is ubiquitous in multistable electronic systems, brown noise is very common in physical and chemical diffusion processes, and even white noise is often encountered in its impulsive from in economics and biological hysteretic systems, so Cauchy or Laplace probability distributions emerge as better model choices in these systems than Gaussian white noise. Various noise models and the numerical methods used in this study to simulate them are discussed in the first part of this chapter. The second part of the chapter is devoted to introducing the theory of stochastic processes defined on graphs recently developed by Freidlin and Wentzell, which proved to be naturally suited to the stochastic analysis of hysteretic systems. First, several definitions and general properties of stochastic processes are discussed, stressing the link between transition probability of Markov processes and the semigroups of contractions. This relationship provides the characterization tool for the diffusion processes that can be defined on a graph, a subject that is elaborated in the final sections of this chapter.

Chapter 3 gives an overview of several common hysteresis phenomena arising in science and engineering. The interest in this topic has been continuously growing over the last years and it has extended far beyond the classical areas of magnetism and plasticity. For example, superconducting hysteresis and economic hysteresis are well-established scientific domains and many pioneering studies have appeared in ecology, wireless communications, psychology, and computer science.

The physical origin of hysteresis is due to the multiplicity of metastable states exhibited by hysteretic systems. A physical system can persist in a metastable state for some time, but thermal perturbations usually drive the system to more stable nearby states. Therefore, the behavior of a hysteretic system could be described as a nonlinear hysteretic transformation of a stochastic input that consists of a random internal noise superimposed on a deterministic external input. In other areas such as economics, computer science, or wireless communication, the external input to a hysteretic system is considered as a stochastic process due to external noise or its random nature. Regardless of the reasons that lead to such models, the systematic study of hysteretic systems driven by stochastic inputs is of relevance to all previously mentioned areas.

Chapter 4 introduces the concept of thermal relaxation, which is a process that refers to the change in the output of a system induced by thermal fluctuations. Examples of thermal relaxation processes can be found in many fields of study, such as magnetism, electronics, or material science. For instance, the magnetization of a ferromagnetic material, the polarization in ferroelectric systems, the binary state of floating gate transistors in flash memories can all change as a result of thermal agitation. The chapter focuses on thermal relaxation processes in scalar and vector hysteretic systems and studies the dynamics of the output variable when the deterministic input is kept constant over time. A few other phenomena related to thermal relaxation such as memory loss and data collapse are also discussed.

Chapter 5 analyzes the spectral density of the output of various hysteretic systems driven by noisy inputs. Closed form analytical solutions for output spectra are derived for bistable hysteretic systems, as well as for complex hysteretic systems that can be described through Preisach model as weighted superposition of symmetric rectangular operators. The mathematical machinery of diffusion processes on graphs is used to circumvent the difficulties related to the non-Markovian property of the output of hysteretic systems. The calculations are appreciably simplified by the introduction of the "effective" distribution function. The implementation of the method for the case of Ornstein-Uhlenbeck process is presented in detail and the general qualitative features of these spectral densities are examined. Due to the universality of the Preisach model, this approach can be used to describe hysteresis nonlinearities of various physical origins. In the last part of this chapter, the spectral density analysis is extended to other models of hysteresis, such as the energetic model, the Jiles-Atherton model, the Coleman-Hodgdon model, and the Bouc-Wen model. The statistical technique for the computation of the output spectra is based on Monte Carlo simulations and Fast Fourier Transforms. The intrinsic differences between the algebraic, differential, and integral modeling of hysteresis are well exposed when the systems are driven by noisy inputs and their stochastic behaviors are compared against each other.

Chapter 6 analyzes the benefits of noise in hysteretic systems by using the framework developed in the previous chapters. While it is mostly seen as a disruptive effect, noise can also have a constructive role by helping a system to overcome a barrier in various activation processes, by providing some degree of randomness useful in audio or visual perceptions, or by activating some kind of resonance response in nonlinear systems. These aspects are introduced and intu-itively explained while providing a short overview of the key results obtained in this area. Although the applications of noise benefits spread over many fields, from climatology and signal processing to nanotechnology and neuroscience, most of the studies can be theoretically framed into two-state models or simple variants thereof, while complex multi-stable systems are rarely addressed. The major contribution of this chapter is to provide a unitary framework for studying con-structive effects of arbitrary colored noise in complex hysteretic systems and its implementation in HysterSoft©. Several examples are discussed following the line of the recent articles published by our group.

Complementary to this book, readers can also download HysterSoft©, which is a software for the simulation of hysteresis and related phenomena (http://www.eng.fsu.edu/ms/HysterSoft). It is a user-friendly simulation framework, in which various mathematical models of hysteresis can be implemented easily. Most numerical results and figures presented in this book have been generated using HysterSoft©, which the authors recommend as a tool for learning and under-standing the concepts discussed in the book. HysterSoft© can also be used to compute first-order reversal-curves (FORC), FORC diagrams, identify the model parameters from experimental data, conduct temperature- and stress-dependent simulations, and analyze the noise passage and stochastic resonance in hysteretic systems. The program can also be used as a dynamic link library (dll) and called from other programs such as Matlab, Simulink, or C++.

The first version of HysterSoft© dates from 2004, when Prof. Petru Andrei worked with Prof. Hans Hauser on hysteresis modeling in magnetic materials at the Technical Institute of Vienna, Austria. The software was very well received by the magnetic hysteresis community and it continued to be improved and expanded over time. In 2011, the software was included in a general finite element simulator for electronic devices-RandFlux©, but it also continued to be distributed as an independent simulation platform on the Florida State University website. The authors are thankful to their collaborators who helped in the development of the software, in particular to Prof. Alexandru Stancu (Al. I. Cuza University, Roma-nia), Prof. David Jiles (Iowa State University, USA), Prof. Amr Adly (Cairo University, Egypt), Prof. Can Korman (Georgetown University, USA), and Dr. Paul Fulmek (Technical University of Vienna, Austria).

Many of the results presented in this book originate from extensive and enlightening discussions with Prof. Isaak Mayergoyz (University of Maryland, USA). The authors are especially grateful to him for his continuous and enthusi-astic support, as well as for his essential role in their professional formation and development. Many thanks are also extended to the professors and colleagues who have been generous in sharing their insights and comments regarding this work, as well as to the graduate students who provided assistance at various stages of the manuscript: Dr. Ayodeji Adedoyin, Dr. Liviu Oniciuc, Dr. Octavian Manu, and Dr. Anca Gindulescu. A special recognition is given to the editors for their dynamic and continuous involvements and encouragements without which this book would not have been possible.

Last, but not least, the authors gratefully acknowledge the financial support from the European Community Seventh Framework Program, the European Social Fund, the US National Science Foundation, the US Army Research Laboratory, and the Romanian Executive Unit for Financing Higher Education and Academic Research.

Suceava, Romania, March 20th, 2013   Dr. Mihai Dimian
Tallahasse, USA, March 20th, 2013   Dr. Petru Andrei

Contents

Chapter 1
Mathematical Models of Hysteresis

1.1 Introduction

Hysteresis is a phenomenon found in many areas of engineering, mechanics, material science, biology, economics, and social sciences. Due to the wide spectrum of areas in which hysteresis is observed and to the fact that the origins of hysteresis are often multiple and unclear, there exist a large number of theoretical models of hysteresis in the literature. It is practically impossible to come up with a single universal model to describe all hysteresis phenomena. So far, most of the existing models of hysteresis were initially developed to describe a particular type of hysteretic system but their mathematical forms were suitable for multi-disciplinary extensions.

In this chapter we present several models of hysteresis that will be used in various applications in subsequent chapters. These models were selected to cover a broad area of studies including mechanics, magnetics, economics, biology, etc.

The **Preisach model** was initially developed to describe the dependence of magnetization on the magnetic field in systems of ferromagnetic particles. Although the model was proposed in the mid 1930s, it became widely used by the scientific community only in the mid 1980s, following the classical works by Mayergoyz [1]. Since then, the model has been extended to describe hysteresis phenomena in many other areas of science such as electromagnetism, economics, biology, and geology, and has become one of the most used mathematical models in the literature. The model has been widely praised for its accuracy in various applications but also criticized for its constraining hypotheses.

The **Bouc-Wen model** originates from an early article of Bouc [2] who proposed a first-order differential equation to describe the loading and unloading curves of the hysteresis loop. The model was subsequently modified by Wen [3] and used mostly for predicting plastic deformations in mechanical systems. The Bouc-Wen model is one of the first and most studied models based on the Duhem hysteresis operator [4] mostly because of its simplicity and to the fact that many of the model properties can be derived analytically.

M. Dimian and P. Andrei, *Noise-Driven Phenomena in Hysteretic Systems*,
Signals and Communication Technology 218, DOI: 10.1007/978-1-4614-1374-5_1,
© Springer Science+Business Media New York 2014

The **Jiles-Atherton model** appeared in the early 1980s and was initially proposed to model hysteresis curves in magnetic materials [5, 6]. The model equations were specifically developed to describe the pinning and rotation of the magnetization in ferromagnetic and ferrimagnetic systems and are related to the physical mechanism of magnetization dynamics in these systems. The model became widely spread particularly after its introduction in SPICE, one of the first and most popular general-purpose analog electronic circuit simulators on the market [7]. In SPICE, the Jiles-Atherton model is used to simulate the magnetic cores of inductors, transformers, and other components containing ferromagnetic or ferritic materials.

The **Coleman-Hodgdon** model appeared in the mid-to-late 1980s as a first-order differential model of hysteresis [8, 9]. The model can be cast in the form of a Duhem hysteresis model; it was initially applied to superconductors and ferrites. Unlike the Jiles-Atherton model, the Coleman-Hodgdon model can be integrated analytically (at least to some degree) so one can derive closed-form expressions for many properties of the hysteresis loops such as saturation, energy loss, differential susceptibilities, etc.

The **energetic model** (also called the Hauser model) appeared in the early 1990s and was primarily used to describe the statistical behavior of magnetic domains under an applied field [10]. The model appears in the form of a transcendental equation for the output variable, which can be solved numerically with relatively little computational overhead. The energetic model has the advantage that it can be easily inverted (i.e. compute the input variable as a function of the output) and implemented numerically. Unlike the Bouc-Wen, Jiles-Atherton, and Colman-Hodgdon models of hysteresis, the energetic model cannot be cast in the form of a Duhem hysteresis model, and the current state of the model depends not only on the current value of the input and output, and the direction of variation (increasing or decreasing) of the input, but also on the past history of the system.

Besides the above models there exists a variety of other hysteresis models in the literature based on either purely mathematical techniques or more physics-based approaches. For instance, the models introduced in [11, 12] are based on Langevin-type approach with positive feedback similar to the effective field method presented in Sect. 1.7.1 of this book. The models introduced in [13–15] are based on the superposition of stop or play operators. The "limiting loop proximity" hysteresis model introduced in [16] presents an algebraic model based on the relative position of the current hysteretic state with respect to the major hysteresis loop. This model is particularly attractive because of its simplicity, since it contains only 4 parameters including the coercive field and the output saturation. The T(x) model introduced in [17] presents a technique based on fitting the hysteresis curves with hyperbolic tangent functions, while the models presented in [18, 19] are based on evolutionary algorithms.

In addition to the mathematical hysteresis models that have little or no physical justification, there are many models based on systems of ordinary or partial differential equations that are derived from a more detailed physical or phenomenological analysis. For instance, the hysteresis induced by charge trapping at the oxide–semiconductor interface in field-effect transistors is usually modeled by solving the semiconductor equations coupled with a trapping model for oxide charges. The hysteresis obtained by charging and discharging lithium batteries is usually modeled by the electron and ion transport equations coupled with Butler-Volmer equations for the reaction rates. Solving these equations usually takes a long simulation time but produces reliable simulation results.

"How to identify what hysteresis model is best fitted for a given application?" This is a question often asked when trying to select a mathematical model to simulate a hysteretic system. In fact, there is no model which is "the best" and, in practice, each model has advantages and disadvantages. For instance, the Preisach model appears to describe the magnetic hysteresis relatively well, but fails to describe the mechanical hysteresis (e.g. plastic deformation of materials, hysteresis induced by friction, etc.) or the hysteresis induced by charging and discharging of energy storage devices, where other models perform much better. Each model is based on an initial set of hypotheses and it is the user's job to verify if these hypotheses hold or not when applied to a given system. In addition, the numerical complexity of hysteresis models varies significantly from model to model, which further limits the area of applicability of the model. For instance, it is relatively easy to use a hysteresis model based on a system of partial differential equations to predict the discharge characteristic of the battery in a laptop because of the high computational power of these devices, but it is almost impossible to use physics-based models in a real-time simulation of transformers with a hysteretic cores.

1.1.1 Definitions, Notations, and Terminology

In this section we introduce some of the terminology used throughout the book when referring to hysteresis phenomena.

In general, by hysteresis we understand a "path-dependent" process, in which the output y depends not only on the current value of the input x, but also on the past values of the input. The state of a hysteretic system can usually be specified at any time by a number of variables called *state variables*. For instance, these variables can be the current values of the input and output, the values of the input and output at the last reversal points, etc. If the values of the state variables are known at time t_0 one can completely describe the behavior of the hysteretic system for any input function $x(t > t_0)$.

Hysteresis may or may not depend on the rate of variation of the input. When it depends on the rate of variation of the input the hysteresis is called *dynamic hysteresis* or *rate-dependent hysteresis*, when it does not depend on the rate of variation of the input the hysteresis is called *static hysteresis* or rate-independent hysteresis. Depending on whether the input and output are scalar or vector variables, we distinguish *scalar* and *vector* hysteretic systems. Unless otherwise noted, in this book we refer to static and scalar hysteretic systems.

The input and output of a hysteretic system can be bounded or unbounded. A hysteretic system is said to be *bounded-input bounded-output* (BIBO) stable if the output variable is bounded for any bounded input variation. Often the output variable is bounded even when the input approaches the boundaries of the interval on which it is defined (for instance to $\pm\infty$). In this case the maximum of the absolute value of the output, y_{sat}, is called *output saturation*.

A *hysteresis curve* is a curve plotted in input–output coordinates. A hysteresis curve in which the input is increasing is called a *loading curve*, while a hysteresis curve in which the input is decreasing is called an *unloading curve*. If the input cycles between two values x_1 and x_2 the output variable traces a *hysteresis loop*, which can be a closed curve or not. Quite often hysteretic systems have a *major hysteresis loop* that is the boundary of the region that encloses all other possible hysteresis curves. If it exists, the major hysteresis loop can often (but not always) be obtained by cycling the input variable between $-\infty$ and $+\infty$. Hysteresis loops that are enclosed in the major hysteresis loop are called *minor loops*.

The input and output of a hysteretic system are often coupled variables. In this case the input is called *generalized force* and the output is called *generalized displacement*. The *energy consumed* when the input goes from x_1 to x_2 is:

$$w = \int_{x_1}^{x_2} y(x)dx. \tag{1.1}$$

If a hysteretic system has a major hysteresis loop, it is customary to define:

(a) The *coercive inputs* $(\pm x_C)$ are the points where the major hysteresis loop crosses the horizontal axis $(y = 0)$. In general, coercive inputs are defined for symmetric major hysteresis loops, with the two critical values equal in magnitude and having opposite sign. Sometimes, we refer to coercive inputs as the *coercive fields* or *coercive forces*. The points where the major hysteresis loop crosses the horizontal axis are called *coercive points*.

(b) The *remanent output* (y_R) is the value of the output for which the major hysteresis loop crosses the vertical axis $(x = 0)$. In general, the remanent output is defined for symmetric major hysteresis loops, with the two remanent values equal in magnitude and having opposite sign. The remanent output is sometimes called *remanent displacement*. The points where the major hysteresis loop crosses the axis $x = 0$ are called *remanent points*.

(c) The *(differential) susceptibility* (χ) is the slope of the hysteresis curve a given point in the input–output plane.

(d) The *anhysteretic curve* $y_{an}(x_0)$ is a curve obtained by representing the final values of the output as a function of x_0, where x_0 is the value of the input obtained as shown in Fig. 1.1. To obtain a point (x_0, y_{an}) on the anhysteretic curve, one applies an alternating series of inputs with infinitesimally slowly decreasing magnitude starting from ∞ and going to 0, and centered around x_0. Since the anhysteretic curve is defined as a limit process, it might not always exist (for instance in the case of the rectangular hysteresis operator defined in Fig. 1.2). If the anhysteretic curve exists, the anhysteretic state obtained for

Fig. 1.1 Input used to measure the *anhysteretic curve*. Notice that the *anhysteretic curve* might not always exist since it is defined as a limit process

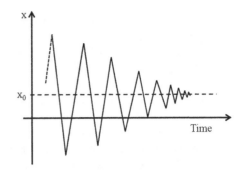

Fig. 1.2 Rectangular hysteretic operator

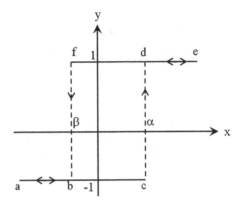

$x_0 = 0$ is often the initial state in many hysteretic computations. This state is called the *zero-anhysteretic state* or *zero-field anhysteretic state*.

(e) The *initial* (or *virgin*) *curve* $y_i(x)$ of a hysteretic system is the curve that is obtained starting from the zero-anhysteretic state and plotting the output as a function of an increasing input. The susceptibility on the initial curve at $x = 0$ is called *initial susceptibility* (χ_i). Notice that the initial curve and the initial susceptibility exist only if the zero-anhysteretic state can be defined.

(f) The differential susceptibilities on the major hysteresis loop at the remanent ($x = 0$ and $y = y_R$) and coercive ($x = x_C$ and $y = 0$) points are called *remanent susceptibility* and *coercive susceptibility* and are denoted by χ_R and χ_C, respectively.

(g) Minor loops are called *closed* if the output variable reaches the initial value after one minor loop. In the case of closed loops, if the input and output are coupled variables, the *energy lost* during a cycle can be computed as

$$w = \int_{x_1}^{x_2} [y_U(x) - y_L(x)]dx = \int_{y_1}^{y_2} [x_U(y) - x_L(y)]dy, \qquad (1.2)$$

where x_1, x_2, y_1, and y_2 are the first and last values of the input and output variables, respectively. $y_L(x)$ and $y_U(x)$ are the equations of the loading and unloading curves in $x - y$ coordinates, and $x_L(y)$ and $x_U(y)$ are the equations of the loading and unloading curves in $y - x$ coordinates. In order for a hysteretic system with coupled input and output variables to be physically possible, the energy lost during a closed hysteresis loop should be positive.

(h) Some hysteretic systems exhibit *accommodation*, which refers to the slight shift of minor hysteresis loops when the input variable varies between the same two values. If the hysteresis loop converges towards a limit hysteresis loop, that loop is called *limit cycle*.

It is often important in applications to study the behavior of a hysteresis model at small values of input variable. In this case the output can usually be expressed as a quadratic function of the input. This is called the law of Lord Rayleigh, who studied this behavior for the first time in magnetic hysteresis [20]. The initial curve can be written as

$$y = \chi_i x \pm bx^2, \tag{1.3}$$

where χ_i is the initial susceptibility and b is a parameter that is usually attributed to irreversible processes. The $+$ sign refers to the loading curve and the $-$ sign to the unloading curve starting from the origin. If the input is cycled between two small values $-x_m$ and x_m, the output can be written as

$$y \pm y_m = \chi_i(x \pm x_m) \pm \frac{b}{2}(x \pm x_m)^2. \tag{1.4}$$

In this case, the output is cycled between $-y_m$ and y_m, where $y_m = \chi_i x_m \pm bx_m^2$. The energy consumed during one cycle can be computed from (1.2), which leads to

$$y = \frac{4}{3}bx_m^3. \tag{1.5}$$

1.1.2 Classification of Hysteresis Models

Due to the large number of hysteresis models in the literature it is rather difficult to perform a comprehensive classification of these models. In this section we provide some insight into the diversity of these models by looking at the type of equations that need to be solved and at the amount of past data that needs to be stored to compute the output variable.

Looking at the type of equations that need to be solved in order to evaluate the output variable we distinguish:

(a) Algebraic models, in which the output is computed by solving algebraic equations;
(b) Differential models, in which the output is computed by solving first or higher order ordinary or partial differential equations;
(c) Integral models, in which the output is given in the form of an integral equation;
(d) Other models in which the output is computed either geometrically or iteratively.

Looking at the amount of past data that needs to be stored to compute the output variable we distinguish:

(a) Duhem-type models of hysteresis, in which the output can change its functional dependence only when the input changes its direction of variation. In these models the future behavior of the hysteretic system can be completely described if the current values of the input and output variables (x, y) and $\text{sign}(\dot{x})$ are known. Mathematically, these models can be expressed as [4]:

$$\dot{y}(t) = f_1(x, y)\dot{x}_+(t) + f_2(x, y)\dot{x}_-(t), \tag{1.6}$$

where $\dot{x}_+(t) = \max(0, \dot{x}(t))$ and $\dot{x}_-(t) = \min(0, \dot{x}(t))$ and the superdot denotes the derivative with respect to time, t. Functions $f_1(x, y)$ and $f_2(x, y)$ can be simple algebraic functions or even differential or integral operators. Many of the existing models of hysteresis are Duhem-type models. For instance the Bouc-Wen, Jiles-Atherton, and Coleman-Hodgdon can all be written in form of (1.6).
(b) Non-Duhem-type models of hysteresis, which take into account not only the coordinates of the last reversal point (x_0, y_0) and the direction of variation of the input, but also other data related to the history of the system. Such models include the Preisach and the energetic models of hysteresis.

Finally, it is worth distinguishing the class of statistical models, in which the output variable is computed as a superposition of elementary hysteresis operators

$$y(t) = \int \int \ldots \int \hat{\gamma}(a_1, a_2, \ldots, a_M) x(t) da_1 da_2 \ldots da_M, \tag{1.7}$$

where $\hat{\gamma}(a_1, a_2, \ldots, a_M)$ is any hysteresis operator that acts on $x(t)$ and depends on some intrinsic parameters $a_1,,\ldots,a_M$ (called model parameters). Examples of such models include the Preisach and a number of models based on the superposition of stop operators.

1.1.3 Parameter Identification Methods

Each hysteresis model contains a number of intrinsic parameters that are usually denoted by a vector π. These parameters need to be determined carefully before using the model. The goal of parameter identification methods (often called parameter determination) is to find an estimate $\hat{\pi}$ of vector π using only measurements of the output variable $y(t)$ as a function of input $x(t)$. According to [21] there are two types of identification methods:

(a) *Recursive methods*, in which the estimate vector $\hat{\pi}$ is found iteratively. Such methods usually start with an initial guess for the model parameters and update that guess iteratively till convergence is obtained. These models are often based on minimization algorithms such as the steepest descends algorithm, the conjugate gradient method, genetic algorithms, or other iterative techniques.
(b) *Nonrecursive methods*, in which the estimate vector $\hat{\pi}$ is found without updating the parameters iteratively. The parameters of the nonrecursive methods are usually directly related to various properties of the input–output characteristics (such as output saturation, coercive field, etc.) and they can be obtained without performing iterations.

It should be mentioned that some identifications methods are using a combination of the above two techniques. Some model parameters are identified directly from the given data, while some other parameters are identified iteratively.

It is also important to mention that the type of the applications where the model is used might require different identification techniques. For instance, in applications where only the major hysteresis loop is important one can determine vector π by fitting the simulated major hysteresis loop to the experimental one without looking at the minor loops. In applications where higher order reversal curves are essential one should use identification methods that take into consideration the shape of the minor loops as well, even if the accuracy of the major loop is somewhat compromised.

Next, we summarize the characteristics of two types of recursive methods that have been used in the literature to compute the parameters of different models of hysteresis. The first type consists in various evolutionary algorithms, while the second one is based on linear least-squares minimization. Both identification techniques are implemented in HysterSoft© and can be used to identify the parameters of all the hysteresis models included in the software. From our experience, both kinds of techniques present a number of advantages and limitations that are discussed below.

1.1.3.1 Identification Techniques Based on Evolutionary Algorithms

Evolutionary algorithms are stochastic search algorithms inspired from biological evolution and/or the social behavior of various species. These algorithms mimic processes from the natural evolution or population systems such as reproduction, mutation, recombination and selection to compute the optimum set of model parameters that fit some experimental data. Next, we discuss a few common features of the existing evolutionary algorithms relevant to the parameter identification of hysteresis models.

Let us denote the model parameters by $\pi = (a_1, a_2, \ldots, a_M)$, where M is the total number of model parameters. For instance, in the case of the energetic model, the model parameters are $\pi = (a, k, q, N_e, c, y_{sat})$, while in the case of the Jiles-Atherton model they are $\pi = (a, \alpha, k, c, y_{sat})$ (see Sects. 1.3 and 1.4 for notations). We also assume that we are given an experimental set of input parameters $p_{exp,i}$ that we can use to compute vector π. Such parameters can be data from the major hysteresis loop, initial curve, or other simple hysteresis curves, such as the coercive force x_C, the remanent value of the output on the major loop y_R, the initial susceptibility χ_i, the susceptibility at coercivity χ_C, the susceptibility at remanence χ_R, etc. If we denote the values of these parameters as predicted by the mathematical model by p_i, the identification problem can be formulated mathematically as a set of equations:

$$p_i(a_1, a_2, \ldots, a_M) = p_{exp,i}, i = 1, \ldots, N, \qquad (1.8)$$

where N denotes the total number of parameters used in the identification problem. If entire output curves are used in the identification problem, these curves can be discretized in a finite number of points (x_i, y_i) and parameters p_i and $p_{exp,i}$ are the simulated and experimental values of the output at each discretization point. Most often, Eq. (1.8) forms a system of highly nonlinear equations that might have a unique solution, multiple solutions, or no solution at all. Hence, instead of attempting to calculate the exact solution of (1.8), most evolutionary algorithms will try to define a *fitness* or *objective* function such as:

$$U(a_1, a_2, \ldots, a_M) = \sum_{i=1}^{N} w_i \left[p_i(a_1, a_2, \ldots, a_M) - p_{exp,i} \right]^2 \qquad (1.9)$$

where w_i are some positive weighting coefficients. If the minimum of function U is zero, then Eq. (1.8) are satisfied exactly. The values of the weighting coefficients depend on the accuracy with which the experimental quantities $p_{exp,i}$ are determined. For instance, if the critical field and the remanent value of the output on the major loop are measured with more accuracy than the susceptibility at coercivity or at the remanent points, the weighting coefficients of x_C and y_R should be larger than the weighting coefficients of χ_C and χ_R. If a parameter is not used in the identification procedure, the corresponding weighting coefficient is set to 0.

Writing the identification problem as the solution of a minimization problem has two major advantages. First, the minimization problem has at least one solution and, second, there exist many algorithms such as the steepest descend method or the conjugate gradient method that can be applied to solve multidimensional minimization problems [9]. Evolutionary algorithms usually require the successive evaluation of the fitness function for different model parameters. The model parameters are changed stochastically from one iteration to another according to the intrinsic laws of the evolutionary model in order to optimize the objective function.

There is a large number of articles in the literature discussing the parameter identification of the Jiles-Atherton, Preisach, and other models of hysteresis based on various metaheuristic optimization algorithms such as genetic algorithms [22–30], shuffled frog leaping algorithm [31], Nelder-Mead method, particle swarm algorithms [32, 33], simulated annealing [31], or other evolutionary algorithms [34]. Unfortunately, the current evolutionary algorithms do not take advantage of the intrinsic formulation of current hysteresis models, which are very nonlinear and whose parameters are subject to complex constraints. For this reason, the objective function should be evaluated many times, which increases the total computation time significantly. In addition, the starting values of model parameters (e.g. the chromosomes in genetic algorithms or the initial population in shuffled frog leaping algorithms) should be close to the optimum solution in order to increase the success rate of the algorithms. From our experience, evolutionary algorithms are computationally much more expensive than other types of identification techniques often diverge, and their applicability is relatively limited.

1.1.3.2 Identification Techniques Based on Linear Least-Square Minimizations

The linear least square minimization technique can be applied to find the model parameters $a_1, a_2, \ldots, a_M$ that minimize the objective function U. Depending on the implementation, linear least square algorithms might require the evaluation of the derivatives of function U with respect to a_j:

$$\partial U(a_1, a_2, \ldots, a_M)/\partial a_j = \sum_{i=1}^{N} 2w_i \left[p_i(a_1, a_2, \ldots, a_M) - p_{\exp,i} \right] \partial p_i(a_1, a_2, \ldots, a_M)/\partial a_j$$

$$(1.10)$$

which, in turn, require the evaluation of the derivatives of parameters p_i with respect to a_j. Since often these derivatives cannot be computed analytically, one can use finite difference approximations to evaluate:

$$\partial p_i(a_1, a_2, \ldots, a_M)/\partial a_j \approx \left[p_i(a_1, \ldots a_j + \varepsilon, \ldots, a_M) - p_i(a_1, \ldots a_j - \varepsilon, \ldots, a_M) \right]/(2\varepsilon)$$

$$(1.11)$$

where ε is a small parameter. The numerical evaluation of the finite differences in (1.11) does not require a large computational overhead because the number of parameters used in the identification problem is relatively small (usually less than ten) and they can be easily evaluated in most models of hysteresis. The computational overhead for the calculation of the minimum of function U on a one-processor personal computer working at 3 GHz is usually less than 2 s for each of the mathematical models used in this book. From our experience, the identification techniques based on least-square minimizations are computationally faster than the evolutionary algorithms, but have worse convergence properties [35]. The initial guess of these techniques should be close (sometimes within a few percentage difference) from the optimum value in order for this algorithm to converge. This limitation of the techniques based on least-square minimizations is due to the strong nonlinearity of hysteresis model equations.

1.1.4 Inverse Modeling

In general, hysteresis models provide a procedure to compute the output variable $y(t)$ as a function of the input variable $x(t)$. In practical applications, it is often necessary to solve the inverse problem, in which one needs to compute the corresponding input variable to a given output variation. This is usually required when solving optimization problems or trying to design a hysteresis system.

Since hysteresis is a nonlinear and multi-valued function, the inverse problem might have no solution, one solution, or multiple solutions. One example is to invert the rectangular hysteresis operator defined in Sect. 1.1.5, which, depending on the output variable, has no solution (if y is different from 1 or -1) or an infinity of solutions (if y is either 1 or -1).

Most often it is possible the invert a hysteresis model relatively easily using the technique described below. Suppose we have a hysteresis model defined as

$$y(t) = \hat{\Gamma}x(t), \qquad (1.12)$$

where $\hat{\Gamma}$ is the hysteresis operator. We introduce the differential susceptibility of the system as

$$\dot{y}(t) = \hat{\chi}\dot{x}(t), \qquad (1.13)$$

where $\hat{\chi}$ is the susceptibility operator. We suppose that such susceptibility operator exists, is finite, and has an inverse operator $\hat{\chi}^{-1}$ such that $\hat{\chi}^{-1}\hat{\chi} = 1$. In this case:

$$\dot{x}(t) = \hat{\chi}^{-1}\dot{y}(t). \qquad (1.14)$$

In the case of scalar hysteresis models $\hat{\chi}$ is the differential susceptibility and $\hat{\chi}^{-1} = 1/\hat{\chi}$. If the hysteresis model is a differential model of hysteresis the computational overhead when solving the direct problem (1.13) is comparable to the computational overhead when solving the inverse problem (1.14).

Most scalar models of hysteresis used in this book can be written in the form of (1.13). Hence, it is possible to invert these models by using (1.14). HysterSoft© implements inverse modeling for all the hysteresis models for which the differential susceptibility is defined, including the Preisach and the energetic models.

1.1.5 Numerical Implementation of Hysteretic Models

Most phenomenological hysteresis models can be expressed in the form of transcendental equations, ordinary differential equations, integral equations, or integro-differential equations. Their numerical implementation can often be done using standard numerical techniques, however, depending on the required accuracy of the final results and on the type of the application, the numerical implementation might require special attention.

From our experience, transcendental equations such as the ones describing the energetic model of hysteresis can be solved using the classical bisection technique. The hysteresis curves are usually monotonically increasing or decreasing so the output value is expected to be found between the last value of the output and $\pm y_{sat}$. The Newton–Raphson technique is usually not appropriate in hysteresis modeling because of the high nonlinearity of the hysteresis curves.

Differential models of hysteresis such as the Jiles-Atheron, energetic, and Bouc-Wen models can usually be integrated using multistep or Runge–Kutta methods with adaptive stepsize. Most of the existing models are not particularly stiff or unstable, which justifies the use of standard integration methods. HysterSoft© uses a fourth-order Runge–Kutta technique with adaptive stepsize to integrate the differential models of hysteresis as well as the rate-dependent models.

Integral models of hysteresis such as the scalar or vector Preisach model can also be integrated using standard adaptive one-dimensional or multi-dimensional integration techniques. It is important to use adaptive techniques since the Preisach distribution function can often be nonzero only on a relatively small region in the Preisach plane and the numerical algorithm might miss this region. Depending on the required accuracy, one might need to use a very refined discretization grid in the region where the Preisach distribution function is nonzero.

Finally, integro-differential models of hysteresis such as dynamic Preisach models, can be integrated either using iterative techniques or by computing the differential susceptibility and integrating it using standard integration methods.

1.1.6 The Rectangular Hysteresis Operator

The rectangular hysteresis operator is one of the simplest hysteresis operators and often stays at the basis of the statistical hysteresis models such as the Preisach model. The memory-based behavior of the output is written as (see Fig. 1.2):

$$y(t) = \hat{\gamma}_{\alpha\beta}x(t) = \begin{cases} 1, & \text{if } x(t) > \alpha \\ 1, & \text{if } x(t) \in [\beta, \alpha] \text{ and } x(t_-) = \alpha \\ -1, & \text{if } x(t) < \beta \\ -1, & \text{if } x(t) \in [\beta, \alpha] \text{ and } x(t_-) = \beta \end{cases} \tag{1.15}$$

where α and β are the "up" and "down" switching values of the input and t_- is the value of the time at the last switching point. If t_- does not exist then we need to specify the initial value of the output variable, which depends on the initial problem that we are modeling. The output of the rectangular hysteresis operator can assume only two values $+1$ and -1. As input $x(t)$ is monotonically increased the ascending branch *abcde* is followed; when the input is monotonically decreased the descending branch *edfba* is followed.

The output of the rectangular hysteresis is BIBO stable and bounded. In addition, there are no hysteresis curves inside or outside the (major) hysteresis loop.

A few other hysteresis operators will be presented in Sect. 1.2.6 when we discuss about the relation of the Preisach model with other hysteresis model. It will be shown that a number of other "elementary" hysteresis operators such as the backslash and elastic–plastic operators can be written as a superposition of rectangular hysteresis operators.

1.2 The Preisach Model

1.2.1 Definition

To introduce the Preisach model we consider an infinite set of rectangular operators $\hat{\gamma}_{\alpha\beta}$ defined in (1.15) and two arbitrary weigh functions $P(\alpha, \beta)$ (with $\alpha > \beta$) and $R(\alpha)$. The output variable is defined as:

$$y(t) = \iint_{\alpha > \beta} \hat{\gamma}_{\alpha\beta} x(t) P(\alpha, \beta) d\alpha d\beta + \int_{-\infty}^{\infty} \text{sign}[x(t) - \alpha] R(\alpha) d\alpha. \qquad (1.16)$$

The first term in the right-hand-side of (1.16) represents purely irreversible switching processes, while the second term represents reversible switching processes. Weight functions $P(\alpha, \beta)$ and $R(\alpha)$ are called the irreversible and reversible components of the Preisach distribution. In this book, by the "Preisach distribution" we will understand in general the set of the two weight functions $P(\alpha, \beta)$ and $R(\alpha)$ (Fig. 1.3). A slightly different definition of the Preisach model was given by Mayergoyz in [36] who allowed the last term in (1.16) to be any real function of x. This approach leads to slightly different formulations of the Preisach model (which are sometimes even easier to write analitically) but is essentially similar to (1.16).

Writing the Preisach model as the superposition of the two components broadens the area of applicability of the model by including non-zero reversal susceptibilities and avoids a number of ambiguities in the identification problem. It has often been proposed in the literature to remove the last term in (1.16) by keeping only the first term and allowing the Preisach distribution to be written in terms of the Dirac-delta function on the diagonal $\alpha = \beta$: $\delta(\alpha - \beta)$. Although apparently simpler, this

Fig. 1.3 Contour plot of the
irreversible component of the
Preisach distribution function
in the Preisach plane. The
units and scale of the
distribution are not shown in
the figure as the Preisach
distribution function is often
normalized to 1

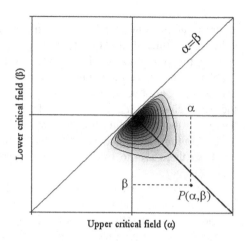

Fig. 1.3 Contour plot of the irreversible component of the Preisach distribution function in the Preisach plane. The units and scale of the distribution are not shown in the figure as the Preisach distribution function is often normalized to 1

approach leads to deceptive mathematical constructions such as $\int_0^\infty \delta(x)\delta x$ and makes the Preisach identification problem ambiguous.

Equation (1.16) can be interpreted geometrically by looking at the Preisach plane, which is obtained by representing $P(\alpha, \beta)$ in the $\alpha - \beta$ coordinates (see Fig. 1.3). If the input traces the path shown in Fig. 1.4a the $\hat{\gamma}_{\alpha\beta}$ operators that are in the -1 state can be separated by the operators in the $+1$ state by the separation line $S(t)$ shown in Fig. 1.4b. This line is often called *staircase line*. All operators to the left and below $S(t)$ (i.e. in region 1) are in the $+1$ state, while the operators to the right and above $S(t)$ (i.e. in region 2) are in the -1 state. The dependence of the output variable on the input is shown in Fig. 1.4c.

1.2.2 General Properties and the Representation Theorem

The following properties can be readily verified for the Preisach model (1.16). Detailed proofs and more explanations can be found in [36].

1. (Wiping-out property) *Only the alternating series of dominant input extrema are stored by the Preisach model. All other input extrema are wiped out.* This property states that the state of the Preisach model is entirely given by the past dominant alternating extrema of the input variation.
2. (Congruency property) *All minor hysteresis loops corresponding to back-and-forth variations of the input between the same two consecutive extreme values are congruent.*

The following representation theorem is due to Mayergoyz [36]: *The wiping-out property and the congruency property constitute the necessary and sufficient conditions for a hysteretic system to be represented by the Preisach model (1.16).*

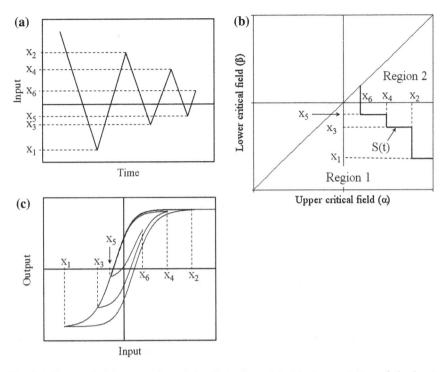

Fig. 1.4 Geometrical interpretation of the Preisach model: (**a**) time variation of the input, (**b**) representation of the Preisach plane (the Preisach distribution is not shown to simplify the figure), and (**c**) the input–output hysteresis curve

This representation theorem is important in practical applications because it gives the conditions under which the Preisach model can be used to simulate a given hysteretic system.

The Preisach distribution can be identified by measuring the first-order reversal curves (FORCs) shown in Fig. 1.5. A FORC can be measured by starting from positive saturation, decreasing the input to a value β, increasing it to α, and measuring the output variables y_β and $y_{\alpha\beta}$. The irreversible component of the Preisach distribution is:

$$P(\alpha, \beta) = \frac{1}{2}\frac{\partial^2 y_{\alpha\beta}}{\partial\alpha\partial\beta}, \tag{1.17}$$

while the reversible component is

$$R(\beta) = \frac{1}{2}\lim_{\alpha\to\beta}\frac{\partial y_{\alpha\beta}}{\partial\alpha}. \tag{1.18}$$

HysterSoft© computes the above derivatives using a special numerical technique based on least-square minimizations initially proposed by Pike [37].

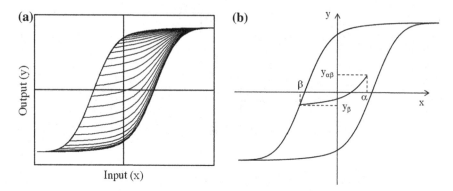

Fig. 1.5 First-order reversal curves used to identify the Preisach distribution. $y_{\alpha\beta}$ is the value of the output after decreasing from positive saturation to $x = \beta$ and then increasing the input to $x = \alpha$

1.2.3 Examples of Preisach Analytical Distributions

Various analytical expressions for the reversible and irreversible components of the Preisach distribution are often being used in the literature. Using analytical expressions for the Preisach distribution has two main advantages: (1) the number of model parameters is significantly reduced to a few fitting parameters simplifying the identification problem considerably, and (2) the double integral in (1.16) can be often computed analytically decreasing the computational overhead of the model. The second column in Table 1.1 shows a few such analytical distributions. The first six examples refer to the reversible component $R(\alpha)$, while the remaining examples to the irreversible component of the distribution $P(\alpha, \beta)$.

In general, the following types of distributions are assumed for $R(\alpha)$ in the literature: uniform, exponential, Gaussian, Cauchy and Langevin distributions, as well as distributions derived from cumulative functions such as Brillouin, tangent hyperbolic, and Langevin functions. It should be noted that the distributions derived from tangent hyperbolic and Langevin function are particular cases of the Brillouin distribution with $J = 1/2$ and $J \to \infty$, respectively.

As for the irreversible component of the Preisach distribution, $P(\alpha, \beta)$, it is customary to assume distributions in terms of the Dirac delta functions, the uniform distribution, or other elementary functions. Line #7 in Table 1.1 shows an example of a distribution in which all the rectangular operators have the same coercivity ξ_0, while line #8 shows an example of a distribution in which all the rectangular operators are symmetric. The non-uniform distributions are usually written either in terms of coordinates α and β, or in terms of ξ and η (defined below) and the Preisach distribution is often written as the product of two terms.

Table 1.1 Everett functions for a few commonly used Preisach functions

ID	Preisach function	Everett function $E(\alpha,\beta)$
#1	$R(\alpha) = \frac{y_{sr}}{2\sigma}\exp\left(-\frac{\|\alpha\|}{\sigma}\right)$	$\frac{y_{sr}}{2}\left[\text{sgn }\beta\exp\left(-\frac{\beta\text{sgn}\beta}{\sigma}\right) - \text{sgn }\alpha\exp\left(-\frac{\alpha\text{sgn }\alpha}{\sigma}\right) + \text{sgn }\alpha - \text{sgn }\beta\right]$
#2[a]	$R(\alpha) = \frac{y_{sr}}{\sqrt{2\pi}\sigma}\exp\left(-\frac{\alpha^2}{2\sigma^2}\right)$	$\frac{y_{sr}}{2}\left[\text{erf}\left(\frac{\alpha}{\sqrt{2}\sigma}\right) - \text{erf}\left(\frac{\beta}{\sqrt{2}\sigma}\right)\right]$
#3[b]	$R(\alpha) = 2y_{sr}\left[-\coth^2\left(\frac{2J+1}{2J}\alpha\right) + \coth^2\left(\frac{1}{2J}\alpha\right)\right]$	$y_{sr}[B_J(\alpha) - B_J(\beta)]$
#4	$R(\alpha) = \frac{y_{sr}}{\sigma}\left[1 - \coth^2\left(\frac{2\alpha}{\sigma}\right) + \frac{\sigma^2}{4\alpha^2}\right]$	$y_{sr}\left[\coth\left(\frac{2\beta}{\sigma}\right) - \coth\left(\frac{2\alpha}{\sigma}\right) - \frac{\sigma}{2\alpha} + \frac{\sigma}{2\beta}\right]$
#5	$R(\alpha) = \frac{y_{sr}}{\sigma}\left[1 - \tanh^2\left(\frac{2\alpha}{\sigma}\right)\right]$	$y_{sr}\left[\tanh\left(\frac{2\alpha}{\sigma}\right) - \tanh\left(\frac{2\beta}{\sigma}\right)\right]$
#6	$R(\alpha) = \frac{y_{sr}}{\pi}\frac{\sigma}{\alpha^2+\sigma^2}$	$\frac{y_{sr}}{\pi}\left[\arctan\left(\frac{\alpha}{\sigma}\right) - \arctan\left(\frac{\beta}{\sigma}\right)\right]$
#7	$P(\alpha,\beta) = \begin{cases} y_{si}\delta(\alpha - \beta - 2\xi_0), & \text{if } \alpha \in [\alpha_1, \alpha_2] \\ 0, & \text{otherwise} \end{cases}$	$\begin{cases} y_{sr}[\min(\alpha_2, \alpha) - \max(\alpha_1, \beta + 2\xi_0)], \\ \quad \text{if } \min(\alpha_2, \alpha) \geq \max(\alpha_1, \beta + 2\xi_0) \\ 0, \text{otherwise} \end{cases}$
#8	$P(\alpha,\beta) = \begin{cases} y_{si}\delta(\alpha + \beta), & \text{if } \alpha \in [\alpha_1, \alpha_2] \\ 0, & \text{otherwise} \end{cases}$	$\begin{cases} y_{sr}[\min(\alpha, -\beta, \alpha_2) - \alpha_1], & \text{if } \alpha \geq \alpha_1 \text{ and } \beta \leq -\alpha_1 \\ 0, \text{otherwise} \end{cases}$
#9	$P(\alpha,\beta) = y_{si}$	$y_{sr}\frac{(\alpha-\beta)^2}{2}$
#10[c]	$P(\alpha,\beta) = \frac{y_{si}}{2\pi\sigma_\alpha\sigma_\beta}\exp\left[-\frac{(\alpha-\alpha_0)^2}{2\sigma_\alpha^2} - \frac{(\beta-\beta_0)^2}{2\sigma_\beta^2}\right]$	$\frac{y_{si}}{2\sqrt{2}\pi\sigma_\alpha}\int_0^\alpha \exp\left[-\frac{(\alpha'-\alpha_0)^2}{2\sigma_\alpha^2}\right]\left[\text{erf}\left(\frac{\alpha'-\beta_0}{\sqrt{2}\sigma_\beta}\right) - \text{erf}\left(\frac{\beta-\beta_0}{\sqrt{2}\sigma_\beta}\right)\right]d\alpha'$
#11[d]	$P(\alpha,\beta) = \frac{y_{si}\sigma}{\pi}g(\eta)\frac{1}{\xi^2+\sigma^2}$	$y_{si}\int_0^{\frac{\alpha-\beta}{\sqrt{2}}}g(s)\left[\arctan\left(\frac{\sqrt{2}\alpha+s}{\sigma_\xi}\right) - \arctan\left(\frac{\sqrt{2}\beta-s}{\sigma_\xi}\right)\right]ds$
#12[d]	$P(\alpha,\beta) = \frac{y_{si}}{\sqrt{2\pi}\sigma_\xi}g(\eta)\exp\left(-\frac{\xi^2}{2\sigma_\xi^2}\right)$	$\frac{y_{si}}{\sqrt{2}}\int_0^{\frac{\alpha-\beta}{2}}g(\sqrt{2}s)\left[\text{erf}\left(\frac{\alpha+s}{\sigma_\xi}\right) - \text{erf}\left(\frac{\beta-s}{\sigma_\xi}\right)\right]ds$

[a] erf is the error function defined as $erf(x) = \frac{2}{\sqrt{\pi}}\int_0^x \exp(-t^2)\,dt$

[b] $B_J(x)$ is the Brillouin function defined as $B_J(x) = \frac{2J+1}{2J}\coth\left(\frac{2J+1}{2J}x\right) - \frac{1}{2J}\coth\left(\frac{x}{2J}\right)$

[c] Usually $\beta_0 = -\alpha_0$ with $\alpha_0 > 0$

[d] ξ and η are functions of α and β and are defined in (1.21) and (1.22).

(a) When written in terms of α and β, the Preisach distribution is often factorized as

$$P(\alpha, \beta) = f(\alpha)g(\beta), \qquad (1.19)$$

where f and g are arbitrary functions of α and β.

(b) When written in terms of ξ and η, the Preisach distribution is often factorized as

$$P(\alpha, \beta) = f(\xi)g(\eta), \qquad (1.20)$$

where

$$\xi = \frac{\alpha + \beta}{\sqrt{2}}, \qquad (1.21)$$

$$\eta = \frac{\alpha - \beta}{\sqrt{2}}. \qquad (1.22)$$

Axis ξ is called the interaction axis and axis η is the coercivity axis. Functions f and g in (1.19) and (1.20) are most often assumed to have Gaussian, Cauchy, or lognormal distributions. For instance, a case which is often encountered in applications is when both $f(\xi)$ and $g(\eta)$ in (1.20) are given by normal distributions:

$$P(\alpha, \beta) = \frac{y_{si}}{2\pi\sigma_\xi\sigma_\eta} \exp\left[-\frac{(\eta - \eta_0)^2}{2\sigma_\eta^2}\right] \exp\left[-\frac{\xi^2}{2\sigma_\xi^2}\right], \qquad (1.23)$$

where y_{si} can be identified as the saturation of the distribution, σ_ξ^2 and σ_η^2 as the variances of the distribution along the two principal axes, and η_0 is the position of the maximum of the distribution along the η axis. Another distribution that we will often use in this book is when $f(\xi)$ is given by a normal distribution while $g(\eta)$ is a lognormal distribution:

$$P(\alpha, \beta) = \frac{y_{si}}{2\pi\sigma_\xi\sigma\eta} \exp\left[-\frac{(\ln\eta - \mu)^2}{2\sigma^2}\right] \exp\left[-\frac{\xi^2}{2\sigma_\xi^2}\right], \qquad (1.24)$$

where σ_ξ^2 is the variance of the distribution along the ξ axis. One can show that the mean of the distribution along the η axis is $\eta_0 = \exp\left(\mu + \frac{\sigma^2}{2}\right)$ and the variance is $\sigma_\eta^2 = [\exp(\sigma^2) - 1]\exp(2\mu + \sigma^2)$.

1.2.4 Computation of Hysteresis Curves in the Preisach Model

Since the evaluation of the double integral in (1.16) requires a large computational cost several techniques have been developed to help the numerical implementation

of the model. One of the most common procedures is to pre-compute the so called
Everett integrals, defined by:

$$E(\alpha, \beta) = \iint_{T(\alpha,\beta)} P(\alpha', \beta')d\alpha'd\beta' + \int_{\beta}^{\alpha} R(\alpha')d\alpha', \text{ for all } \alpha \geq \beta, \quad (1.25)$$

where triangle T is represented in Fig. 1.6. If these integrals are pre-computed and
stored at all reversal points (x_i, x_{i+1}) [see Fig. 1.4b], the output variable can be
computed as:

$$y = -E(x_0, x_0) + 2 \sum_{1}^{n-1} E(x_i, x_{i+i}) \quad (1.26)$$

if the first extreme value of the input x_0 was a maximum. If the first extreme value
of the input x_0 was a minimum, the output variable can be computed as:

$$y = E(x_0, x_0) - 2 \sum_{1}^{n-1} E(x_i, x_{i+i}) \quad (1.27)$$

where, in both equations, n is the current number of extreme points stored by the
model.

It is apparent from the above analysis that the Preisach model can be given in
terms of either the Preisach distribution functions $P(\alpha, \beta)$ and $R(\alpha)$ or the Everett
integral $E(\alpha, \beta)$. In fact, in many practical applications it might be more advan-
tageous to know the values of the Everett integral $E(\alpha, \beta)$ instead of the Preisach
distribution functions. The third column in Table 1.1 shows the analytical
expressions of the Everett function for the Preisach distributions presented in the
second column. These functions are also implemented in HysterSoft©.

Next, we summarize the analytical equations for the most important hysteresis
curves and discuss about the initial susceptibility and the susceptibility at reversal points.

Fig. 1.6 Triangle in the
Preisach plane where the
Everett integral is computed

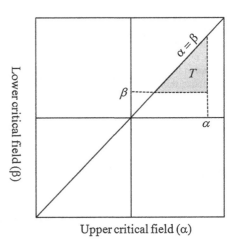

Lower critical field (β)

Upper critical field (α)

(a) *Anhysteretic curve*. If the anhysteretic curve exists (this usually happens if the Preisach distribution function does not have singularities), the output can be evaluated as:

$$y_{an}(x) = \iint_{A^+(x)} P(\alpha, \beta)d\alpha d\beta - \iint_{A^-(x)} P(\alpha, \beta)d\alpha d\beta$$
$$+ \int_{-\infty}^{x} R(\alpha)d\alpha - \int_{x}^{\infty} R(\alpha)d\alpha \tag{1.28}$$

In the case when the Preisach distribution is symmetric with respect to the $\beta = -\alpha$ axis the anhysteretic curve can be written as

$$y_{an}(x) = 2\int_{0}^{\infty} d\eta \int_{0}^{\sqrt{2}x} P\left(\frac{\xi + \eta}{\sqrt{2}}, \frac{\xi - \eta}{\sqrt{2}}\right)d\xi + 2\int_{0}^{x} R(\xi)d\xi, \tag{1.29}$$

where ξ and η are defined in (1.21) and (1.22) (Fig. 1.7). The second column in Table 1.2 summarizes the anhysteretic functions corresponding to the Preisach distributions from Table 1.1.

(b) *Initial curve*. The initial hysteresis curve is defined by starting with the system in the zero-anhysteretic state. The initial curve can be written as:

$$y_i(x) = E(\infty, -\infty) + 2\int_{-\infty}^{\infty} d\beta' \int_{\beta'}^{\max(x, -\beta')} P(\alpha', \beta')d\alpha'$$
$$+ 2\int_{-\infty}^{x} R(\alpha')d\alpha' \tag{1.30}$$

where, in the case of hysteretic systems with saturation $E(\infty, -\infty) = y_{sat}$. In the case when the Preisach distribution is symmetric with respect to the $\beta = -\alpha$ axis the initial curve can be written as $y_i(x) = E(x, -x)$.

Fig. 1.7 Preisach plane used to compute the anhysteretic curve. A^+ denotes the area below the anhysteretic separation line and A^- the area above this line

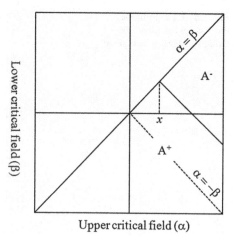

Lower critical field (β)

Upper critical field (α)

Table 1.2 The anhysteretic functions, initial susceptibilities, output saturation, and distribution parameters for the Preisach distribution functions shown in Table 1.1

ID	Anhysteretic curve	Initial susceptibility (χ_i)	Output saturation (y_{sat})	Distribution parameters
#1	$y_{sr}\left[1 - \exp\left(-\frac{|x|}{\sigma}\right)\right]$	$\frac{y_{sr}}{\sigma}$	y_{sr}	y_{sr}, σ
#2	$y_{sr}\,\mathrm{erf}\left(\frac{x}{\sqrt{2}\sigma}\right)$	$\sqrt{\frac{2}{\pi}}\frac{y_{sr}}{\sigma}$	y_{sr}	y_{sr}, σ
#3	$y_{sr}\left[\frac{2J+1}{2J}\coth\left(\frac{2J+1}{2J}x\right) - \frac{1}{2J}\coth\left(\frac{1}{2J}x\right)\right]$	∞	y_{sr}	y_{sr}, J
#4	$y_{sr}\left[\coth\left(\frac{2x}{\sigma}\right) - \frac{\sigma}{2x}\right]$	$\frac{y_{sr}}{3\sigma}$	y_{sr}	y_{sr}, σ
#5	$y_{sr}\tanh\left(\frac{2x}{\sigma}\right)$	$\frac{2y_{sr}}{\sigma}$	y_{sr}	y_{sr}, σ
#6	$\frac{2y_{sr}}{\pi}\arctan\left(\frac{x}{\sigma}\right)$	$\frac{2y_{sr}}{\pi\sigma}$	y_{sr}	y_{sr}, σ
#7	$\begin{cases} y_{si}\delta(\alpha - \beta - 2\xi_0), & \text{if } \alpha \in [\alpha_1, \alpha_2] \\ 0, & \text{otherwise} \end{cases}$	0		$y_{si}, \alpha_1, \alpha_2, \xi_0$
#8	$\begin{cases} 0, & \text{if } x = 0 \\ \mathrm{sgn}(x)y_{si}(\alpha_2 - \alpha_1), & \text{otherwise} \end{cases}$	0		$y_{si}, \alpha_1,$
#9	$\begin{cases} 0, & \text{if } x = 0 \\ \infty, & \text{if } x > 0 \\ -\infty, & \text{if } x < 0 \end{cases}$	0	∞	y_{si}
#10	Eq. (1.28)	0	$\leq y_n$	$y_{si}, \alpha_0, \beta_0, \xi_0, \sigma_\alpha, \sigma_\beta$
#11	$\frac{2y_{si}}{\pi}\arctan\left(\frac{x}{\sigma}\right)$	0	$y_{si}\int_0^\infty g(s)ds$	$y_{si}, \sigma, g(\eta)$
#12	$y_{si}\mathrm{erf}\left(\frac{x}{\sqrt{2}\sigma}\right)$	0	$y_{si}\int_0^\infty g(s)ds$	$y_{si}, \sigma, g(\eta)$

(c) *Loading curve.* The loading major hysteresis curve starts from $x = -\infty$ and can be written in terms of the Everett integral (1.25) as:

$$y_L(x) = -E(\infty, -\infty) + 2E(x, -\infty), \tag{1.31}$$

where, in the case of hysteretic systems with saturation $E(\infty, -\infty) = y_{sat}$.

(d) *Unloading curve.* The unloading major hysteresis curve starts from $x = \infty$ and can be written in terms of the Everett integral (1.25) as:

$$y_U(x) = E(\infty, -\infty) - 2E(x, +\infty). \tag{1.32}$$

(e) *Initial susceptibility.* The initial susceptibility in the Preisach model can be computed by taking the derivative of (1.30) with respect to x when $x \to 0$. If the irreversible component of the Preisach distribution does not have a singularity at $(\alpha, \beta) = (0,0)$ the only contribution to the initial susceptibility comes from the reversible component:

$$\chi_i = 2R(0). \tag{1.33}$$

The third column in Table 1.2 summarizes the initial susceptibilities corresponding to the Preisach distributions from Table 1.1.

(f) *Susceptibility at reversal points.* Let x_r be the value of the input at a reversal point. If the irreversible component of the Preisach distribution does not have a singularity in the limit $(\alpha, \beta) \rightarrow (x_r, x_r)$ the susceptibility at reversal point x_r is:

$$\chi_i = 2R(x_r). \tag{1.34}$$

The last equation can be used to identify the reversible component of the distribution, $R(x)$.

1.2.5 Generalizations of the Preisach Model

The Preisach model was subject to many modifications over the years. These modifications were made in order to extend the area of applicability of the model to other physical (particularly magnetic) systems that do not satisfy the wiping-out or the congruency properties of the classical Preisach model. In this section we summarize a few such generalizations that will appear later in the book.

1.2.5.1 The Moving Preisach Model

The moving Preisach model is given by (1.16) in which the input variable in the right-hand side is changed to an effective input given by

$$x_{eff} = x + f(x, y). \tag{1.35}$$

The above model exhibits the congruency and wiping-out properties in the operative plane define by variables (x_{eff}, y).

1.2.5.2 The Input-Dependent Preisach model

The input-dependent Preisach model is obtained using input dependent Preisach distribution functions $P(\alpha, \beta, x)$ and $R(\alpha, x)$

$$y(t) = \iint_{\alpha \geq \beta} \hat{\gamma}_{\alpha\beta} x(t) P(\alpha, \beta, x) d\alpha d\beta + \int_{-\infty}^{\infty} \hat{\gamma}_{\alpha\alpha} x(t) R(\alpha, x) d\alpha. \tag{1.36}$$

The previous equation can be written in a slightly different form due to Mayergoyz [36], which simplifies the parameter identification technique:

$$y(t) = \iint_{R} \hat{\gamma}_{\alpha\beta} x(t) P(\alpha, \beta, x) d\alpha d\beta + \frac{y_L + y_U}{2}, \tag{1.37}$$

where rectangle R is defined by $\alpha[$ and β (see Fig. 1.8a) while y_L and y_U are the loading and unloading major hysteresis curves, respectively. Equation (1.37) can be derived from (1.36) by splitting integration domain in the first integral into R and triangles T_1 and T_2, and observing that the integrals over T_1 and T_2 combined with the last term in (1.36) leads to average of the loading and unloading major hysteresis curves. Notice that, according to definition (1.37) the irreversible Preisach distribution function needs to be defined only for $b < x < a$.

The input dependent Preisach model has the following properties:

1. (Wiping-out property) *Only the alternating series of dominant input extrema are stored by the Preisach model. All other input extrema are wiped out.*
2. (Congruency property) *All minor hysteresis loops corresponding to back-and-forth variations of the input between the same two consecutive extreme values have equal vertical chords.*

The following representation theorem is due to Mayergoyz [36]: *The above wiping-out property and the congruency properties constitute the necessary and sufficient conditions for a hysteretic system to be represented by the input-dependent Preisach model* (1.36).

The Preisach distribution of the input-dependent Preisach model can be identified by measuring the second-order reversal curves (SORC) shown in Fig. 1.18. A SFORC is measured by starting from positive saturation, decreasing the input to a value β, increasing it to $\alpha > \beta$, decreasing it to x, and measuring the output variable $y_{\alpha\beta x}$. The irreversible and reversible components of the Preisach distribution are can be determined as (Fig. 1.8b)

$$P(\alpha, \beta, x) = \frac{1}{2}\frac{\partial^2 y_{\alpha\beta x}}{\partial\alpha\partial\beta}, \text{ if } \beta < x < \alpha \tag{1.38}$$

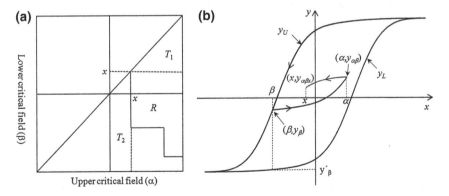

Fig. 1.8 The Preisach plane and ad second-order reversal curve (SORC) used to compute the input-dependent Preisach distribution

The condition that the first-order ascending and descending curves are symmetric imply that the Preisach distribution functions need to satisfy the following equations

$$P(\alpha, \beta, x) = P(-\alpha, -\beta, -x) \text{ and } R(\beta, x) = R(-\beta, -x) \qquad (1.39)$$

The Preisach distribution of the input-dependent Preisach model can be computed in HysterSoft© if a set of SORCs are provided.

1.2.5.3 Other Preisach-Type Models

A variety of other Preisach-type models have been developed in the literature. For instance, the restricted Preisach-model introduced in [36] has been developed to relax the congruency property of the model to the so called "comparable" minor loops. A number of other modes were introduced by Della Torre [38–51] in order to relax the congruency property even more and account for various effects such as accommodation and memory loss. In these models, the output is computed iteratively since Preisach distribution often depends on the current value of the output variable. Another model with a variable Preisach distribution was introduced in [52]; this model was shown to be more accurate for particulate ferromagnetic media, particularly for systems of highly interactive magnetic particles.

Another type of Preisach model was introduced by Roshko et al. [53, 54] and improved by Stancu et al. [55]. This model was used by a number of research groups to describe temperature dependent hysteresis in systems of magnetic particles.

1.2.6 Relation Between the Preisach Model and Other Models of Hysteresis

A number of other hysteresis models can be directly related to or are special cases of the Preisach model. Below we present three such models that appeared more frequently in the literature.

1.2.6.1 Relay Hysteresis Operator

The relay hysteresis operator can be regarded as a generalization of the rectangular hysteresis operator in the sense that the output can trace two fixed output curves $h_U(x)$ or $h_L(x)$ according to (see Fig. 1.9)

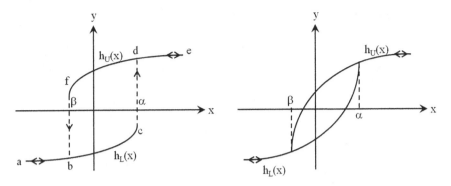

Fig. 1.9 Examples of relay hysteretic operators

$$y(t) = \begin{cases} h_U(x(t)), & \text{if } x(t) \geq \alpha \\ h_U(x(t)), & \text{if } x(t) \in (\beta, \alpha) \text{ and } x(t_-) = \alpha \\ h_L(x(t)), & \text{if } x(t) \leq \beta \\ h_L(x(t)), & \text{if } x(t) \in (\beta, \alpha) \text{ and } x(t_-) = \beta \end{cases} \tag{1.40}$$

where α, β, and t_- have the same significance as in (1.15). If no previous switching exists, the value of the output is either $h_U(x(t))$ or $h_L(x(t))$ depending on the initial hysteretic state of the system. Depending on the choice of functions $h_U(x(t))$ and $h_L(x(t))$, the output of the relay hysteresis operator can be BIBO stable or not. In addition, there are no hysteresis curves inside or outside the major hysteresis loop defined by functions $h_U(x(t))$ and $h_L(x(t))$.

The relay hysteretic operator satisfies the wiping-out and congruency properties of the input-dependent Preisach model and, hence, can be written in terms of the Preisach model. The input-dependent Preisach distribution can be identified as:

$$P\left(\tilde{\alpha}, \tilde{\beta}, x\right) = \frac{1}{2}[h_U(x) - h_L(x)]\delta(\tilde{\alpha} - \alpha)\delta\left(\tilde{\beta} - \beta\right), \tag{1.41}$$

$$R(\tilde{\alpha}, x) = \frac{1}{2}[h_U(x) + h_L(x)]\delta(\tilde{\alpha} - \alpha)\text{sgn}(x - \tilde{\alpha}). \tag{1.42}$$

1.2.6.2 The Backlash (Play) Operator

The backlash operator (also called the play or the Krasnosel'skii-Pokrovskii operator), is defined by

$$y(t) = \max[x(t) - h, \min(x(t) + h, y(t_-))] \tag{1.43}$$

where h is the coercivity of the hysteretic loop and t_- is the time when the last input extreme was attained. The hysteretic loop is represented in Fig. 1.10a. It is usually assumed that initial state of the operator is $x = y = 0$. The backlash

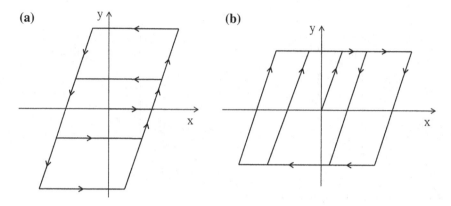

Fig. 1.10 The backlash (**a**) and the stop (**b**) operators

operator is a particular case of the Preisach model, in which the Preisach distribution is given by

$$P(\alpha, \beta) = y_{si}\delta(\alpha - \beta - h) \text{ and } R(\alpha) = 0. \tag{1.44}$$

1.2.6.3 The Stop Operator

The stop operator (also called the elasto-plastic or Prandtl operator) is defined by

$$y(t) = \min[h, \max(-h, x(t) - x(t_-) + y(t_-))] \tag{1.45}$$

where t_- denotes that last value of the input reversal [see Fig. 1.10b]. It is usually assumed that initial state of the operator is $x = x_0 = y = 0$. The stop operator is the dual of the backlash operator and can also be regarded as a particular case of the Preisach model, in which the Preisach distribution is given by

$$P(\alpha, \beta) = y_{sat}\delta(\alpha - \beta - h) \text{ and } R(\alpha) = y_{sat}\delta(\alpha - \beta). \tag{1.46}$$

The stop operator stays at the basis of a number of models based on superposition [56–58]. However, all these models are particular cases of the Preisach model of hysteresis.

1.3 Jiles-Atherton Model

1.3.1 Definition

The Jiles-Atherton model of hysteresis describes the output variable by the following differential equation [5, 6]:

$$\frac{dy}{dx} = (1-c)\,\delta\,\frac{L(x+\alpha y) - y}{k(1-c)\,\text{sgn}(\dot{x}) - \alpha[L(x+\alpha y) - y]} + c\,\frac{dL(x+\alpha y)}{dx}, \quad (1.47)$$

where c, α, and k are the model parameters, which are assumed to be non-negative, and L is the anhysteretic curve. Although this curve can have different shapes depending on the type of each hysteretic system, so far, it was mostly assumed to be a Langevin function:

$$L(x) = y_{sat}\left[\coth\left(\frac{x}{a}\right) - \frac{a}{x}\right], \quad (1.48)$$

where a is another fitting parameter and y_{sat} is the saturation value of the output. In Eq. (1.47) δ is a parameter that was equal to 1 in the original versions of the Jiles-Atherton model [5, 6] (note that δ was also referring to sign $(d\dot{x}/dt)$ in the original model), but, was latter set to [59]:

$$\delta = \begin{cases} 0, & \text{if } \dot{x}[L(x+\alpha y) - y] \le 0 \\ 1, & \text{otherwise} \end{cases} \quad (1.49)$$

in order to avoid curves with negative differential susceptibility (see Fig. 1.11). HysterSoft© distinguishes between the two cases by setting variable Version to either Y1992_Jiles when $\delta = 1$ or to Y1994_Deane when Eq. (1.49) is used.

The following constraints have to be satisfied by the anhysteretic function in order for the model to be BIBO stable:

(a) $k(1-c)\,\text{sgn}(\dot{x}) > \alpha[L(x+\alpha y) - y]$ for all possible values of x and y.

(b) $1 - \alpha c\frac{dL(x)}{dx} > 0$ for any x. If the anhysteretic function is given by (1.48) this condition is equivalent to $3a > \alpha c y_{sat}$.

So far, the model has been mostly applied to magnetic hysteresis and introduced in circuit simulators to model nonlinear inductors and transformers [7, 60]. In magnetic hysteresis, parameter α accounts for the demagnetization energy, k for domain

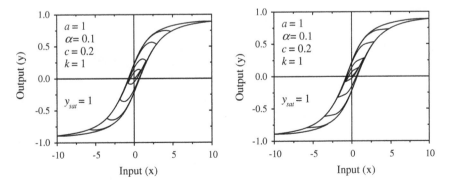

Fig. 1.11 Minor and major loops simulated using the Jiles-Atherton model with $\delta = 1$ (*left*) and δ given by Eq. (1.49) (*right*)

pinning, and parameter c is introduced to separate explicitly the reversible and irreversible components of the magnetization. More information about the derivation and physical significance of the model parameters can be found in [5, 6, 61, 62].

1.3.2 General Properties of Hysteretic Curves

It can be shown that the Jiles-Atherton model has the following properties:

(a) The output variable is bounded and lies in interval $(-y_{sat}, y_{sat})$.
(b) The hysteretic state of the Jiles-Atherton model is completely described by the values of the input, output, and the direction (increasing or decreasing) of the input variable.
(c) The model displays accommodation effects. If the input is cycled between two values the hysteresis curves tend towards a limit cycle (see Fig. 1.12).
(d) The differential susceptibility is non-negative when Eq. (1.49) is satisfied but it can be positive, negative, or 0 when $\delta = 1$.
(e) Equation (1.47) can be written in the form of a nonlinear first-order differential equation by taking the derivative in the last term in (1.47) and using the chain rule. One obtains:

$$\frac{dy}{dx} = \frac{(1-c)\,\delta\,[L(x+\alpha y)-y]}{k(1-c)\,\mathrm{sgn}(\dot{x}) - \alpha[L(x+\alpha y)-y]} + c\,\frac{dL(x_{eff})}{d(x_{eff})}\left(1 + \alpha\frac{dy}{dx}\right), \qquad (1.50)$$

and after a few rearrangements:

$$\frac{dy}{dx} = \frac{(1-c)\,\delta\,\dfrac{L(x+\alpha y)-y}{k(1-c)\,\mathrm{sgn}(\dot{x})-\alpha[L(x+\alpha y)-y]} + c\,\dfrac{dL(x_{eff})}{dx_{eff}}}{1 - \alpha c\,\dfrac{dL(x_{eff})}{dx_{eff}}}, \qquad (1.51)$$

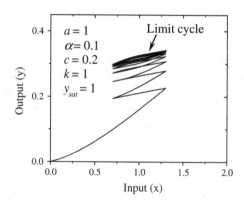

Fig. 1.12 Accommodation effects and limit cycle in the Jiles-Atherton model

where

$$x_{eff} = x + \alpha y, \qquad (1.52)$$

is the "effective" value of the input

(f) The anhysteretic curve is given by function L [see Eq. (1.48)]. This fact can be proved by using the fact that the model has a unique limit cycle and taking the limit of closed small cycles around the anhysteretic function (similar to the procedure used to find the anhysteretic curve of the Coleman-Hodgdon model in Sect. 1.6.2).

(g) If the anhysteretic function L is given by (1.48) the slope of the anhysteretic function at $x = 0$ is

$$\chi_{an} = \frac{\left.\dfrac{dL(x_{eff})}{dx_{eff}}\right|_{x_{eff}=0}}{1 - \alpha \left.\dfrac{dL(x_{eff})}{dx_{eff}}\right|_{x_{eff}=0}} = \frac{y_{sat}}{3a - \alpha y_{sat}}. \qquad (1.53)$$

(h) The initial susceptibility can be computed by imposing $\mathrm{sgn}(\dot{x}) = 1$, $x = 0$ and $y = 0$ in (1.51)

$$\chi_i = \frac{c\dfrac{dL(x_{eff})}{dx_{eff}}}{1 - \alpha c\dfrac{dL(x_{eff})}{d(x_{eff})}} = \frac{c y_{sat}}{3a - \alpha c y_{sat}}. \qquad (1.54)$$

where, to derive the last term in (1.54) we have assumed again that anhysteretic function L is given by (1.48).

(i) The susceptibility at the critical fields can be computed by imposing $\mathrm{sgn}(\dot{x}) = 1$, $x = x_C$ and $y = 0$ in (1.51) where x_C is the coercive field. One obtains:

$$\chi_C = \frac{(1 - c)\dfrac{L(x_C)}{k(1-c) - \alpha L(x_C)} + c\dfrac{dL(x_C)}{dx_C}}{1 - \alpha c\dfrac{dL(x_C)}{dx_C}}. \qquad (1.55)$$

In most applications $\frac{dL(x_C)}{dx_C}$ can be neglected in Eq. (1.55), which leads to a slightly simpler expression for χ_C.

(j) The coefficient of the quadratic term that enters into the law of Lord Rayleigh (1.4) can be expresses as a function of the model's parameters as [63]:

$$b = \frac{a^2 y_{sat}(1 - c)}{2k(a - c\alpha y_{sat})^3}. \qquad (1.56)$$

The coefficient of the linear term in the law of Lord Rayleigh is of course given by the initial susceptibility.

1.3.3 Parameter Identification Methods

The existing parameter identification methods for the Jiles-Atherton model are usually iterative techniques based on least-square minimizations, genetic algorithms, or non-iterative techniques. The non-iterative identification methods are usually based on the following observations:

(1) The output saturation is given by model parameter y_{sat}.
(2) The coercive field of the model is mostly dependent on parameter k. The effect of the other model parameters on the value of the coercive field is usually much smaller.
(3) The initial and zero-field anhysteretic susceptibilities [Eqs. (1.54) and (1.53), respectively] do not depend on parameter k.
(4) The effective input x_{eff} is relatively close to the real input x so the last term in (1.52) can often be neglected.
(5) The α parameter does not have a significant effect on the value of the coercive field but changes the remanence point and the differential susceptibilities at the coercive point significantly.

Next, we summarize an identification method due to Jiles et al. [64] that finds the model parameters by fitting the initial curve, the anhysteretic curve, and various parameters from the major hysteresis loop to experimental data. The equations below can be derived directly from the constitutive equation of the model (1.47) or from the general properties of the hysteretic curves presented in the previous subsection.

It is assumed that the anhysteretic function is given by (1.48) and the output saturation is already measured experimentally and equal to parameter y_{sat}. Hence, we only need to determine parameters a, α, c, and k. The algorithm proceeds as follows:

(1) One starts with initial guess for parameter $\alpha = 0$.
(2) Estimate parameter c using the values of the initial and zero-field anhysteretic susceptibilities:

$$c = \frac{\chi_i}{\chi_{an}} \frac{3a - \alpha y_{sat}}{3a - \alpha c y_{sat}} \tag{1.57}$$

(this equation can be obtained from (1.53) and (1.54)).
(3) Estimate parameter a from the anhysteretic susceptibility at $x = 0$:

$$a = \frac{y_{sat}}{2}\left(\frac{1}{\chi_{an}} + \alpha\right). \tag{1.58}$$

(4) Compute parameter k from the value of the coercive field x_C and the susceptibility at the coercive field χ_C:

$$k = \frac{L(x_C)}{1-c}\left[\alpha + \frac{1-c}{\chi_C - c\frac{dL(x_{eff})}{dx_{eff}}\Big|_{x_e=x_C}}\right].$$ (1.59)

(5) Compute parameter α from the value of the output y_R and the differential susceptibility at remanence χ_R by solving the following transcendental equation:

$$\chi_R = -(1-c)\frac{L(\alpha y_R) - y_R}{k(1-c) + \alpha[L(\alpha y_R) - y_R]} + c\frac{dL(x_{eff})}{d(x_{eff})}\Big|_{x_e=\alpha y_R}(1 + \alpha\chi_R)$$ (1.60)

(6) Repeat steps (3)–(5) till convergence is attaint.
(7) Go to step (2) and repeat the algorithm till the value of parameter c does not change significantly.

This identification technique is also implemented in HysterSoft©.

1.4 Energetic (Hauser) Model

1.4.1 Definition

The output variable in the energetic (or the Hauser) model is equal to [10, 65, 66]:

$$y = -y_{sat}, \text{ if } x \le x_{sat,1},$$ (1.61)

$$y = y_{sat}, \text{ if } x \ge x_{sat,2},$$ (1.62)

and is computed by solving the following transcendental equation:

$$x = N_e y + \text{sgn}(y)hG(y) + \text{sgn}(y - y_0)\left[\frac{k}{y_{sat}} + c_r hG(y)\right]$$
$$\times \left[1 - \kappa\exp\left(-\frac{q}{\kappa y_{sat}}|y - y_0|\right)\right], \text{ if } x_{sat,1} \le x \le x_{sat,2},$$ (1.63)

where

$$G(y) = G_{sat}\left(\frac{y}{y_{sat}}\right),$$ (1.64)

$$G_{sat}(s) = \left[(1+s)^{1+s}(1-s)^{1-s}\right]^{\frac{g}{2}} - 1,$$ (1.65)

$$x_{sat,1} = -N_e y_{sat} - h(2^g - 1) - \left[\frac{k}{y_{sat}} + c_r h(2^g - 1)\right]$$
$$\times \left[1 - \kappa \exp\left(-\frac{q(y_{sat} + y_0)}{\kappa y_{sat}}\right)\right] \tag{1.66}$$

$$x_{sat,2} = N_e y_{sat} + h(2^g - 1) + \left[\frac{k}{y_{sat}} + c_r h(2^g - 1)\right]$$
$$\times \left[1 - \kappa \exp\left(-\frac{q(y_{sat} - y_0)}{\kappa y_{sat}}\right)\right] \tag{1.67}$$

In the above equations y_{sat}, N_e, h, g, c_r, k, and q are positive model parameters and y_{sat} can be identified as the output saturation. y_0 is the value of the output at the last reversal point. Initially, the simulations start with $x = 0$ and $y_0 = 0$ and, then, the output variable is computed using (1.61)–(1.63).

The major hysteresis loop and sample reversal curves are shown in Fig. 1.13. In these simulations the major loop saturates at $x_{sat,1} = x_{sat,2} \approx 42.2$. The energetic model predicts a "closure" point of the loading and unloading branches of the major loop. The major loop saturates (i.e. $y = \pm y_{sat}$) at these "closure" points.

In the literature parameter k is usually normalized by the vacuum permeability μ_0 ($= 1.256 \times 10^{-6}$ H/m), so parameter k should be replaced by k/μ_0 when comparing the results obtained this section with the ones in the literature. In this book we would like to extend the area of applicability of the model to non-magnetic systems, for which reason we define $\mu_0 = 1$. HysterSoft© allows users to set the vacuum permeability parameter to 1 or μ_0, depending on the preference.

Parameter κ depends on the past history of the system and, at each reversal point, it is re-computed as:

$$\kappa = 2 - \kappa_0 \exp\left(-\frac{q}{y_{sat}\kappa_0}|y - y_0|\right), \tag{1.68}$$

Fig. 1.13 The major hysteresis loop (*dash line*) and sample hysteresis curves (*continuous line*) simulated with the energetic model of hysteresis

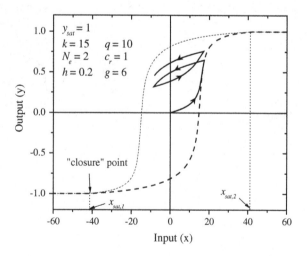

where κ_0 and y_0 are the values of κ and y computed at the last reversal point. The simulation always starts with the initial curve ($y_0 = 0$ and $\kappa = 1$). At each reversal point κ is calculated by (1.68) and y_0 is set to the actual value of the output at this reversal point. Parameter κ varies between 0 and 2.

Notice that Eq. (1.68) is equivalent to imposing that the output variable does not have discontinuities at reversal points. Indeed, if we consider reversal point b with coordinates (x_0, y_0), point c with coordinates (x_1, y_1) (see Fig. 1.14), and take the limit $y \to y_0$, Eq. (1.63) gives

$$x_1 = N_e y_1 + hG(y_1) + \left[\frac{k}{y_{sat}} + c_r hG(y_1) \right] \times \left[1 - \kappa_0 \exp\left(-\frac{q}{\kappa_0 y_{sat}} |y_1 - y_0| \right) \right]$$

$$(1.69)$$

in the case of curve b-c and

$$x_1 = N_e y_1 + hG(y_1) - \left[\frac{k}{y_{sat}} + c_r hG(y_1) \right] \times (1 - \kappa) \qquad (1.70)$$

in the case of curve c-d. By subtracting the last two equations we obtain (1.68), where $y \equiv y_1$.

The energetic model has often been used in the literature under the following two approximations:

$$\exp\left(-\frac{q}{\kappa} \right) \approx 0, \qquad (1.71)$$

$$2^g \gg 1. \qquad (1.72)$$

In order to keep the generality of the model, we do not use the above approximations in the following sections. However, many of the properties of the hysteretic curves presented in the following can be simplified by using (1.71) and (1.72).

Fig. 1.14 Continuity of hysteresis curves at reversal points

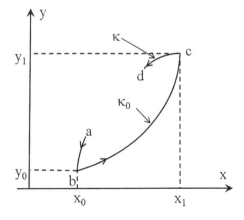

1.4.2 General Properties of Hysteretic Curves

The energetic model has the following properties:

(a) The output variable is bounded and lies in the interval $(-y_{sat}, y_{sat})$.
(b) The hysteretic state of the energetic model is completely described by the values of the input, output, parameter κ, and direction (increasing or decreasing) of the input variable.
(c) The model displays accommodation. However, if the input varies alternatively between the same 2 extreme values, say x_0 and x_1 there exist a limit cycle between points (x_0, y_0) and (x_1, y_1), which can be computed from (1.61)–(1.68). The κ parameter converges towards $\kappa_{\text{limit cicle}}$, which is the solution of the following equation

$$\kappa_{\text{limit cicle}} = 2 - \kappa_{\text{limit cicle}} \exp\left(-\frac{q}{y_{sat}\kappa_{\text{limit cicle}}}|y_2 - y_1|\right). \tag{1.73}$$

When the limit cycle extends from $-y_{sat}$ to y_{sat} we obtain the major hysteresis loop and parameter κ converges towards κ_∞ defined by

$$\kappa_\infty = 2 - \kappa_\infty \exp\left(-\frac{2q}{\kappa_\infty}\right). \tag{1.74}$$

The mapping $q \to \kappa_\infty$ is a bijection (see Fig. 1.15) and κ_∞ increases monotonically from 1 to 2 when parameter q is varied from 0 to ∞.

(d) The initial hysteresis curve (for $x > 0$) is given by

$$x = N_e y_i + hG(y_i) + \left[\frac{k}{y_{sat}} + c_r hG(y_i)\right] \times \left[1 - \exp\left(-\frac{qy_i}{y_{sat}}\right)\right]. \tag{1.75}$$

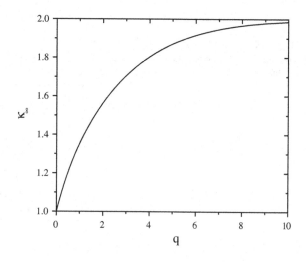

Fig. 1.15 Parameter κ_∞ as a function of q on the major hysteresis loop

(e) The major hysteresis loop of the energetic loop cannot be computed by starting from the initial state with $\kappa = 1$, saturating the system in one direction and, then, in the other direction, because parameter κ does not have a unique value in the limit $x \to \pm\infty$ so the "loading" and "unloading" curves that start from $\pm\infty$ depend on the past history of the model. Hence, for the purpose of the presentation we define the major loop of the energetic model as the limit cycle when x is cycled between $\pm\infty$. The equations of the loading y_L and unloading y_U curves of the major hysteresis loop (defined as mentioned) can be written as:

$$x = N_e y_L + \operatorname{sgn}(y_L) h G(y_L)$$
$$+ \left[\frac{k}{y_{sat}} + c_r h G(y_L) \right] \times \left[1 - \kappa_\infty \exp\left(-\frac{q(y_{sat} + y_L)}{\kappa_\infty y_{sat}} \right) \right] \quad (1.76)$$

$$x = N_e y_U + \operatorname{sgn}(y_U) h G(y_U)$$
$$- \left[\frac{k}{y_{sat}} + c_r h G(y_U) \right] \times \left[1 - \kappa_\infty \exp\left(-\frac{q(y_{sat} - y_U)}{\kappa_\infty y_{sat}} \right) \right] \quad (1.77)$$

If approximations (1.71) and (1.72) hold parameter $\kappa \approx 2$ when $x \to \pm\infty$ and the major hysteresis loop as defined in Sect. 1.1.1 is identical to the limit cycle defined above. If these approximations do not hold, the hysteresis curves can cross and go out of the limit cycle defined by (1.76) and (1.77).

(f) The differential susceptibility of the model is non-negative and can be expressed in terms of the model parameters as follows

$$\frac{dy}{dx} = 0, \text{ if } x \leq x_{sat,1} \text{ or } x \geq x_{sat,2}, \quad (1.78)$$

and

$$\left(\frac{dy}{dx} \right)^{-1} = N_e + h\operatorname{sgn}(y)G'(y) + \frac{q}{\kappa y_{sat}} \left[\frac{k}{y_{sat}} + c_r h G(y) \right] \exp\left(-\frac{q}{\kappa y_{sat}} |y - y_0| \right)$$
$$+ \operatorname{sgn}(y - y_0) c_r h \left[1 - \kappa \exp\left(-\frac{q}{\kappa y_{sat}} |y - y_0| \right) \right] G'(y),$$

$$\quad (1.79)$$

if $x_{sat,1} \leq x \leq x_{sat,2}$, where $G'(y)$ is the derivative of function $G(y)$.

(g) Since $G(0) = G'(0) = 0$ (see Fig. 1.16) the differential susceptibility at the coercive points (on the major hysteresis loop) becomes

$$\chi_C = \frac{1}{N_e + \frac{qk}{y_{sat}^2} \exp\left(-\frac{q}{\kappa_\infty} \right)}. \quad (1.80)$$

If approximation (1.71) is satisfied, the differential susceptibility at the

Fig. 1.16 Function $G_{sat}(s)$ defined in (1.65)

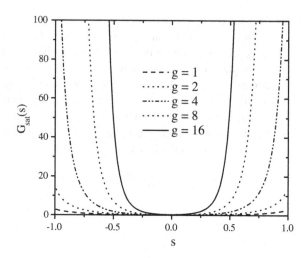

coercive becomes $\chi_C \approx \frac{1}{N_e}$.

(h) The initial susceptibility is:

$$\chi_i = \frac{1}{N_e + \frac{qk}{y_{sat}^2}} \tag{1.81}$$

(i) The coercive field is:

$$x_C = \frac{k}{y_{sat}} \left[1 - \kappa_\infty \exp\left(-\frac{q}{\kappa_\infty}\right) \right]. \tag{1.82}$$

If approximation (1.71) is satisfied, the differential susceptibility at the coercivity points becomes $x_C \approx \frac{k}{y_{sat}}$.

(j) The energy lost during one cycle of the major hysteresis loop can be computed by subtracting Eqs. (1.76) and (1.77) and using (1.1):

$$w = k \left[4 - \frac{8}{q}(e^{-q} - 1) \right]$$

$$+ 2c_r h \int_{-y_{sat}}^{y_{sat}} G(y) \left[1 - \kappa_\infty \exp\left(-\frac{q}{\kappa_\infty}\right) \cosh\left(-\frac{qy}{\kappa_\infty y_{sat}}\right) \right] dy \tag{1.83}$$

If approximation (1.71) is satisfied, the integral can be evaluated numerically to obtain:

$$w \approx 4k + 2.359 c_r h. \tag{1.84}$$

(k) The analytical expression for the remanent output can be computed by setting $x = 0$ in (1.63) and solving numerically the resulting transcendental equation.

(l) The coefficient of the quadratic term that enters into the law of Lord Rayleigh (1.4) can be expresses as a function of the model's parameters as:

$$b = \frac{y_{sat}^3}{2} \frac{kq^2 - hg y_{sat}}{kq + y_{sat}^2 N_e}.$$
(1.85)

The last expression can be derived by considering y_i is a function of x in (1.75) and expending expression (1.75) in Taylor series.

1.4.3 Parameter Identification Methods

Next we present a parameter identification method for the energetic model of hysteresis. The presentation follows closely the method presented in [67] but deviates significantly in the way the various parameters on the major hysteresis loop are used. The parameter identification method presented below aims to determine the six model parameters by using the following data (see Fig. 1.17):

- x_C, the coercive field,
- χ_i, the initial susceptibility,
- χ_C, the differential susceptibility at coercivity,
- y_R, the remanent output on the major loop,
- x_R, the input corresponding to output y_R on the loading curve.
- (x_g, y_g), an arbitrary point on the unloading curve of the major loop.

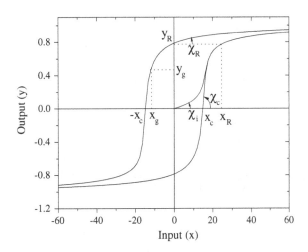

Fig. 1.17 Physical parameters used to identify the parameters of the energetic model (k, q, N_e, h, g, and c_r)

These variables can be expressed as functions of the six model parameters by using the following system of equations that can be derived from the model's basic equations:

$$x_C = \frac{k}{y_{sat}}\left(1 - \kappa_\infty e^{-\frac{q}{\kappa_\infty}}\right), \tag{1.86}$$

$$\frac{1}{\chi_i} = N_e + \frac{qk}{y_{sat}^2}, \tag{1.87}$$

$$\frac{1}{\chi_c} = N_e + \frac{qk}{y_{sat}^2}e^{-\frac{q}{\kappa_\infty}}, \tag{1.88}$$

$$N_e y_R + hG(y_R) - F(-y_R) = 0, \tag{1.89}$$

$$N_e y_g + hG(y_g) + F(y_g) = x_g, \tag{1.90}$$

$$N_e y_R + hG(y_R) + F(y_R) = x_R, \tag{1.91}$$

where κ_∞ is given by (1.74) and

$$F(y) = F_{sat}\left(\frac{y}{y_{sat}}\right), \tag{1.92}$$

$$F_{sat}(s) = \left[\frac{k}{y_{sat}} + c_r hG_{sat}(s)\right]\left\{1 - \kappa_\infty \exp\left[-\frac{q}{\kappa_\infty}(1+s)\right]\right\}, \tag{1.93}$$

and $G_{sat}(s)$ is given by (1.65). Equations (1.86)–(1.91) represent a system of six nonlinear equations that need to be solved for the six model parameters. All model parameters should be positive in order for the model to describe hysteresis with positive differential susceptibilities. The conditions under which Eqs. (1.86)–(1.91) have no solution, one solution, or multiple solutions are analyzed next. For now, let us note that Eqs. (1.86)–(1.88) involve only parameters k, q, N_e so these three equations can be solved first. Hence, the following subsection focuses on the computation of k, q, N_e alone; after that, we present the technique for the computation of the remaining parameters.

1.4.3.1 Computation of k, q, and N_e

Equations (1.87) and (1.88) can be rearranged to obtain the following equation that needs to be solved numerically for q:

$$\left(\frac{1}{\chi_i} - \frac{1}{\chi_c}\right)\frac{y_{sat}}{x_c} = q\frac{1 - e^{-\frac{q}{\kappa_\infty(q)}}}{1 - \kappa_\infty(q)e^{-\frac{q}{\kappa_\infty(q)}}}. \tag{1.94}$$

$\kappa_\infty(q)$ is given by solving (1.74). Function

$$f(q) = q \frac{1 - e^{-\frac{q}{\kappa_\infty(q)}}}{1 - \kappa_\infty(q)e^{-\frac{q}{\kappa_\infty(q)}}} \tag{1.95}$$

has a minimum of 2 at $q_{min} = 0$ (see Fig. 1.18). Hence, Eq. (1.94) implies $\left(\frac{1}{\chi_i} - \frac{1}{\chi_c}\right)\frac{y_{sat}}{x_c} \geq f(q_{min}) = 2$, which shows that the space of physical parameters that the energetic model can describe is constrained by the inequality:

$$\frac{1}{\chi_i} \geq \frac{1}{\chi_c} + \frac{2x_c}{y_{sat}}. \tag{1.96}$$

Initial susceptibilities that do not specify this inequality cannot be modeled by the energetic model of hysteresis.

Once Eq. (1.94) is solved for q, parameters k and N_e can be found from:

$$k = \frac{y_{sat}x_c}{1 - \kappa_\infty e^{-\frac{q}{\kappa_\infty}}}, \tag{1.97}$$

$$N_e = \frac{1}{\chi_i} - \frac{qk}{y_{sat}^2}. \tag{1.98}$$

Note that, as long as inequality (1.96) is satisfied, Eq. (1.94) will always have unique solution, which can be computed by using the standard bisection technique. It is often important in applications to find the conditions that the physical parameters should satisfy in order for N_e (which corresponds to the demagnetizing factor in magnetic materials) to be positive. This condition can be derived by plugging (1.97) into (1.98) and imposing $N_e > 0$. The space of possible physical parameters for a hysteretic system with $x_C = 15$ and $y_{sat} = 1$ is represented in

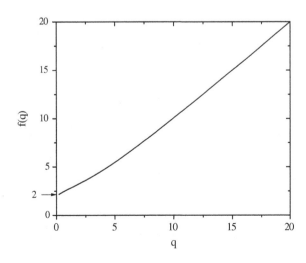

Fig. 1.18 Function $f(q)$ defined in (1.95)

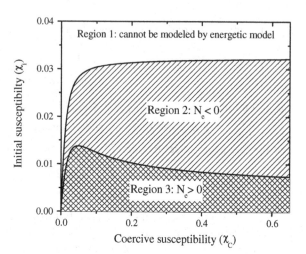

Fig. 1.19 Only initial and coercive susceptibilities given by the points in regions 2 and 3 can be modeled by the energetic model in a hysteretic system with $x_C = 15$ and $y_{sat} = 1$. For relatively large values of χ_i the model predicts a negative demagnetizing factors N_e as implied by Eq. (1.98). Initial and coercive susceptibilities given by the points in region 1 cannot be modeled by the model

Fig. 1.19 by regions 2 and 3. For a system with an initial susceptibility of 7×10^{-3} and susceptibility at coercivity of 0.3, we obtain the following set of model parameters by solving (1.94), (1.97), and (1.98): $q = 9.2$, $k = 15.36$, and $N_e = 1.92$.

In the next subsections we describe the technique for the computation of parameters h, g, and $c_r = 0$. We consider two cases that appear often in the literature: (1) the case when $c_r = 0$, in which the modeled is called the classical energetic model, and (2) the case when $c_r > 0$, which gives the complete energetic model.

1.4.3.2 Computation of h and g in the Classical Energetic Model $(c_r = 0)$

In the case of the classical energetic model $c_r = 0$ and Eqs. (1.89)–(1.91) simplify substantially. By substituting h from (1.89) into (1.90) we obtain the following equation that needs to be solved numerically for g:

$$x_g = N_e y_g + \frac{G(y_g)}{G(y_R)} \left[\frac{k}{y_{sat}} Q(y_R) - N_e y_{sat} y_R \right] + \frac{k}{\mu_0 M_s} Q(y_g), \qquad (1.99)$$

where:

$$Q(y) = Q_{sat}\left(\frac{y}{y_{sat}}\right), \qquad (1.100)$$

$$Q_{sat}(s) = 1 - \kappa_\infty \exp\left[-\frac{q}{\kappa_\infty}(1+s)\right]. \qquad (1.101)$$

Once parameter g is computed, parameter h can be found by solving Eq. (1.89). We obtain:

$$h = \frac{k}{y_{sat}} \frac{Q(-y_R)}{G(y_R)} - \frac{N_e y_{sat} y_R}{G(y_R)}. \tag{1.102}$$

Equations (1.94), (1.97)–(1.102) should be solved iteratively in order to compute the model parameters. For instance, in the case of a hysteretic system with $y_{sat} = 1$, $x_C = 15$, $\chi_i = 6.6 \times 10^{-3}$ and $\chi_C = 0.33$ we obtain $h = 0.2$ and $g = 9$.

It is important to analyze now the conditions under which $h > 0$ and $g > 0$. The condition $h > 0$ implies that y_R should be smaller than some limiting value $y_{R,\max}$, which can be found by solving $\frac{k}{y_{sat}} Q(-y_{R,\max}) > N_e y_{R,\max}$. Due to the monotonicity of functions G and Q, if $y_R < y_{R,\max}$ Eq. (1.102) will always have unique solution. The condition $g > 0$, on the other hand, gives a limited set of values for x_g for any fixed y_g. After some algebraic manipulations one can distinguish two cases:

(1) If $y_R < y_{R,\max}$ then x_g should satisfy $x_g > N_e y_g + hG(y_g)\big|_{g=0} + F(y_g)$, which provides a lower limit for the possible values of the applied fields for which magnetization can be y_g.
(2) If $y_R > y_{R,\max}$ then x_g is bounded from both above and below as follows

$$N_e y_g + \frac{k}{y_{sat}} Q(y_g) > x_g > N_e y_g + hG(y_g)\big|_{g=0} + F(y_g), \tag{1.103}$$

These conditions are important for solving the identification problem because hysteretic systems that do not satisfy them cannot be modeled by the energetic model. For instance, in the case of a hysteretic system with $x_C = 15$, $y_{sat} = 1$, $\chi_i = 6.6 \times 10^{-3}$, $\chi_C = 0.33$ and $y_g = 0.835$, the space of available physical parameters is given by region 1 in Fig. 1.20.

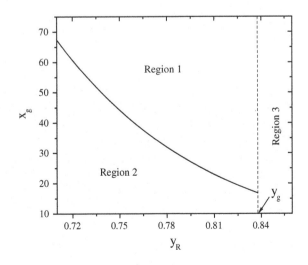

Fig. 1.20 Possible values of input x_g corresponding to an output of $y_g = 0.835$. Only parameters (x_g, y_R) in region 1 can be described by the classical energetic model

1.4.3.3 Computation of h, g, and c_r in the Complete Energetic Model

In the case of the complete energetic model (i.e. when $c_r > 0$) and parameters g, h, and c_r can be computed from the following equations:

$$x_g = N_e y_g + \frac{G(y_g)}{G(y_R)} + \frac{k}{y_{sat}} Q(y_g) + I(y_R, y_g),$$
(1.104)

$$h = \frac{1}{G(y_R)} \left[\frac{x_R Q(-y_R)}{Q(y_R) + Q(-y_R)} - N_e y_R \right],$$
(1.105)

$$c_r = \frac{1}{hG(y_R)} \left[\frac{H_r}{Q(y_R) + Q(-y_R)} - \frac{k}{y_R} \right],$$
(1.106)

which can be obtained from (1.89)–(1.91). In the last equations we introduced notation

$$I(y_R, y_g) = x_R \frac{Q(-y_R) + Q(y_g)}{Q(y_R) + Q(-y_R)} - \frac{k}{y_{sat}} Q(y_g) - N_e y_R.$$
(1.107)

Eq. (1.104) is a transcendental equation in g and should be solved numerically. By using the same line of reasoning as in the previous subsection we can derive the conditions for which h and g are positive. Condition $h > 0$ implies that

$$\frac{(x_R - N_e y_R)Q(-y_R) - N_e y_R Q(y_R)}{Q(y_R) + Q(-y_R)} > 0,$$
(1.108)

which shows that y_R should lie outside the interval $\left[y_{R,\min}, y_{R,\max}\right]$, where $y_{R,\min}$ and $y_{R,\max}$ are the roots of the numerator and denominator in (1.108), respectively. It can be shown that both the numerator and the denominator in (1.108) have unique solution in interval $[0, y_{sat}]$, which confirms the existence of the limits of interval $\left[y_{R,\min}, y_{R,\max}\right]$.

The possible values of (y_R, y_g) that can be described by the complete energetic model can be found by imposing $g > 0$ in Eq. (1.104). It can be shown again that the possible values of x_g that correspond to a measured value of the output equal to y_g depend on whether y_g is larger or smaller than y_R. Equations somewhat more complex than the ones derived in the previous subsection hold in this case as well: if $y_R < y_g$, x_g should be larger than some minimum value that can be computed by setting $g = 0$ and solving (1.104) for x_g; if $x_R > x_g$, x_g is found between two critical values that can be computed by solving (1.104) for x_g in the limits $g = 0$ and $g = \infty$. Figure 1.21 shows the space of available physical parameters for a hysteretic system with $x_C = 15$, $y_{sat} = 1$, $\chi_i = 6.6 \times 10^{-3}$, $\chi_C = 0.33$ and $y_g = 0.8$, in the $x_R - x_g$ plane. Applying Eqs. (1.104)–(1.106) we obtain $h = 0.217$, $g = 8.24$, and $c_r = 1.32$.

Fig. 1.21 Possible values of input x_g corresponding to an output of $y_g = 0.8$. Only parameters (x_g, y_R) in regions 2 and 4 can be described by the classical energetic model

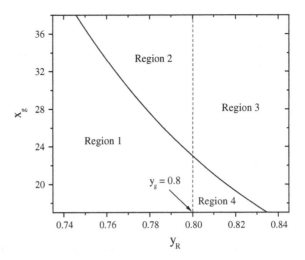

1.5 Bouc-Wen Model

1.5.1 Definition

The output variable satisfies the following equation:

$$y = \alpha kx + (1 - \alpha)Dkz, \tag{1.109}$$

where z is a function of the input variable x and its derivative $\dot{x}$, α ($0 \le \alpha \le 1$) and k are two model parameters. Function z is usually given in terms of a differential equation. Table 1.3 presents a few examples used in the literature for the differential equation that function z can satisfy. It is customary to refer to the last model on the table as the Bouc-Wen model.

1.5.2 General Properties of Hysteretic Curves

A detail description of various parameter identification methods for the Bouc-Wen model can be found in [21]. Below we summarize a few general properties of the model.

(a) Depending on the choice of model's parameters the Bouc-Wen system can be BIBO stable or BIBO unstable. A classification of the cases in which the Bouc-Wen model is BIBO stable is given in Table 1.4 where the following notations are used

Table 1.3 Summary of Bouc models of hysteresis. The Bouc-Wen model of hysteresis is generally referred to model (1.113)

Reference	$\dot{z}$	
Bouc [2]	$\dot{z} = D^{-1}[A\dot{x} - \beta z|\dot{x}|]$	(1.110)
Bouc [2]	$\dot{z} = D^{-1}[A\dot{x} - \beta\,\mathrm{sgn}(z)|\dot{x}| - \gamma\dot{x}]$	(1.111)
Wen [3]	$\dot{z} = D^{-1}[A\dot{x} - \beta\,\mathrm{sgn}(z)|\dot{x}||z|^n - \gamma\dot{x}|z|^n]$	(1.112)
Baber-Noori [68–70]	$\dot{z} = \frac{h(z)}{\eta}[A\dot{x} - \beta\,\mathrm{sgn}(z)|\dot{x}||z|^n - \gamma\dot{x}|z|^n]$	(1.113)

$$z_0 = \sqrt[n]{\frac{A}{\gamma+\beta}} \text{ and } z_1 = \sqrt[n]{\frac{A}{\gamma-\beta}} \qquad (1.114)$$

The domain of BIBO stability denotes the domain of the initial condition (z_0) of the differential Eq. (1.113) for which the model is BIBO stable.

(b) The Bouc-Wen model has a finite output saturation if $\alpha = 0$. In this case the output saturation can be computed from the last column in Table 1.4.

(c) If input $x(t)$ is periodic and cycling between values x_1 and x_2 the output converges uniformly to a continuous function (see Fig. 1.22).

(d) The hysteresis curves obtained by plotting $y(x)$ are traced in clockwise direction. Hence, variable x and y cannot represent a pair of conjugate variables such as generalized forces and generalized displacements because the system would produce energy. However, the Bouc-Wen model can be used to describe correctly other mechanical variables that lead to energy dissipation (see Chap. 3).

The Bouc-Wen model does not have a major hysteresis loop and the "coercive field" increases to infinity when $x \to \pm\infty$.

(e) The differential susceptibility of the model is

$$\frac{dy}{dx} = \alpha k + (1-\alpha)k[A - \beta\mathrm{sgn}(z\dot{x})|z|^n - \gamma|z|^n], \qquad (1.115)$$

where $z = \frac{y-\alpha kx}{(1-\alpha)k}$.

(f) The initial susceptibility is

$$\chi_i = \alpha k + (1-\alpha)kA. \qquad (1.116)$$

Table 1.4 Classification of BIBO stable Bouc-Wen model parameters [21]

Case		Domain of BIBO stability for z_0	Upper bound of z(t)
$A > 0$	$\beta+\gamma>0$ and $\beta-\gamma\geq0$	$(-\infty,\infty)$	$\max(|z(0)|,z_0)$
	$\beta-\gamma<0$ and $\beta\geq0$	$[-z_1,z_1]$	$\max(|z(0)|,z_0)$
$A < 0$	$\beta-\gamma>0$ and $\beta+\gamma\geq0$	$(-\infty,\infty)$	$\max(|z(0)|,z_1)$
	$\beta+\gamma<0$ and $\beta\geq0$	$[-z_0,z_0]$	$\max(|z(0)|,z_1)$
$A = 0$	$\beta+\gamma>0$ and $\beta-\gamma\geq0$	$(-\infty,\infty)$	$|z(0)|$

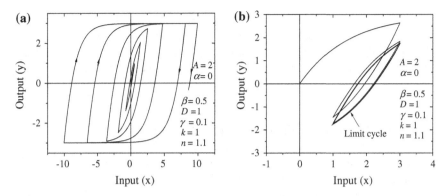

Fig. 1.22 Clockwise hysteresis curves (**a**) and the limit cycle (**b**) simulated by the Bouc-Wen model

1.6 Coleman-Hodgdon Model

1.6.1 Definition

There are two versions of the Coleman-Hodgdon model that have been used to model static hysteresis in the literature. The first version of the model has been applied mostly to superconductors [71], while the second version to magnetic materials [72, 73]. The constituent equations of both versions are presented in this section.

1.6.1.1 The Direct Coleman-Hodgdon Model

In the direct Coleman-Hodgdon model the output variable satisfies the following differential equation:

$$\frac{dy}{dx} = \alpha(x)\text{sgn}(\dot{x}) \cdot [f(x) - y] + g(x), \qquad (1.117)$$

where f, g, and α are some piecewise smooth material functions.

1.6.1.2 The Reverse Coleman-Hodgdon Model

In the reverse Coleman-Hodgdon model the output variable satisfies the following differential equation:

$$\frac{dy}{dx} = \frac{1}{\alpha(y)\text{sgn}(\dot{y}) \cdot [f(y) - x] + g(y)}, \qquad (1.118)$$

where f, g, and α are some piecewise smooth material functions. Notice that, although we have used the same notations, the material functions in (1.118) have a different significance than the ones in (1.117). All the equations that are derived from model (1.118) are also valid for (1.117) by interchanging the input and output variables.

1.6.1.3 Conditions on the Choice of Material Functions

Any piecewise smooth material functions in (1.117) and (1.118) leads to some kind of hysteresis of the output variable. Below we enumerate a few constraints that that these functions should satisfy in order to be able to use them in super-conductor and magnetic systems. Then, we present a few standard functions that satisfy these constraints and are often used in the literature:

i. f should be a piecewise continuous and even function that increases mono-tonically, has a derivative with a finite limit $f(\infty)$ for large x, and

$$f(0) = 0. \tag{1.119}$$

ii. g should be a piecewise continuous, even function with

$$\lim_{x \to \infty} g(x) = f'(\infty). \tag{1.120}$$

and

$$f'(x) \geq g(x), \tag{1.121}$$

for all x, in the case of the direct model (1.117), and

$$f'(x) \leq g(x), \tag{1.122}$$

for all x, in the case of the reverse model (1.118)

iii. α should be a piecewise continuous, even function of x satisfying

$$\alpha(x) > 0, \text{ for all } x \tag{1.123}$$

and

$$\lim_{x \to \pm\infty} \alpha(x) > 0, \text{ for all } x. \tag{1.124}$$

Conditions (1.121) and (1.122) guarantee that the hysteresis loops are traced in counterclockwise direction. Condition (1.120) ensures the closure of the major loop.

A standard choice for the material functions of the direct Coleman-Hodgdon model used to model hysteresis in superconductor materials is [71, 74]:

$$f(x) = \begin{cases} A_1 \arctan(A_2 x), \text{ if } |x| \le x_{bp} \\ A_1 \arctan(A_2 x_{bp}) + x - x_{bp}, \text{ if } x > x_{bp} \\ -A_1 \arctan(A_2 x_{bp}) + x + x_{bp}, \text{ if } x < -x_{bp} \end{cases} \tag{1.125}$$

$$g(x) = \begin{cases} f'(x) \cdot \left[1 - A_3 \exp\left(-\frac{A_4|x|}{x_{cl}-|x|}\right)\right], \text{ if } |x| < x_{cl} \\ f'(x), \text{ if } |x|_{cl} \end{cases} \tag{1.126}$$

$$\alpha(x) = \alpha_0 \operatorname{sech}\left(\frac{x}{\alpha_1}\right) + \alpha_2 \tag{1.127}$$

where A_1, A_2, A_3, A_4, x_{bp}, x_{cl}, α_0, α_1, and α_2 are some model parameters.

Model (1.118) seems to be more appropriate for magnetic hysteresis, where one can assume that $\alpha(y) = \alpha_0$ (is constant) and [75]:

$$f(y) = \begin{cases} D_2 y, \text{ if } |y| \le y_{bp} \\ D_1(y - y_{cl}) + D_2 y_{cl}, \text{ if } y > y_{bp} \\ D_1(y + y_{cl}) - D_2 y_{cl}, \text{ if } y < -y_{bp} \end{cases} \tag{1.128}$$

$$g(y) = \begin{cases} (1 + D_3)D_2, \text{ if } |y| < y_{cl} \\ D_1, \text{ if } |y| \ge y_{cl} \end{cases} \tag{1.129}$$

$$\alpha(y) = \alpha_0 \tag{1.130}$$

where α_0, D_1, D_2, D_3, y_{bp}, and y_{cl} are some model parameters. Another choice appropriate for magnetic hysteresis described by model (1.118) is given below:

$$f(y) = \begin{cases} A_1 \tan(A_2 y), \text{ if } |y| \le y_{bp} \\ A_1 \tan(A_2 y_{bp}) + (y - y_{bp})/\mu_s, \text{ if } y > y_{bp} \\ -A_1 \tan(A_2 y_{bp}) + (y + y_{bp})/\mu_s, \text{ if } y < -y_{bp} \end{cases} \tag{1.131}$$

$$g(y) = \begin{cases} f'(y) \cdot \left[1 - A_3 \exp\left(-\frac{A_4|y|}{y_{cl}-|y|}\right)\right], \text{ if } |y| < y_{cl} \\ f'(y), \text{ if } y \ge |y_{cl}| \end{cases} \tag{1.132}$$

where μ_s, A_1, A_2, A_3, A_4, y_{bp}, and y_{cl} are some model parameters.

1.6.2 General Properties of Hysteretic Curves

In this section we present the general properties of the Coleman-Hodgdon hysteresis model (1.118). The properties presented below can be translated to model (1.117) by interchanging the input and output variables in each equation. These properties can be derived from Eq. (1.118). The derivation of some of these properties can be found in [71] and [76].

(a) The future behavior of the model is completely described by the current hysteretic state given by the values of the input variable, output variable, and direction (increasing or decreasing) of the input variable. Hence, the Coleman-Hodgdon model is a Duhem-type model of hysteresis (see Sect. 1.1.2).

(b) The initial susceptibility is:

$$\chi_i = \frac{1}{g(0)}. \tag{1.133}$$

(c) The Coleman-Hodgdon model has two solutions, one corresponding to the loading (L) and the other one to the unloading (U) curves, for each initial point (x_0, y_0). These solutions can be computed by integrating (1.118):

$$x_L(y|x_0, y_0) = f(y) + [x_0 - f(y_0)]e^{-\tilde{\alpha}(y|y_0)} + \int_{y_0}^{y} h(s)e^{\tilde{\alpha}(s|y)}ds, \text{ for } y \geq y_0 \tag{1.134}$$

$$x_U(y|x_0, y_0) = f(y) + [x_0 - f(y_0)]e^{\tilde{\alpha}(y|y_0)} - \int_{y}^{y_0} h(s)e^{-\tilde{\alpha}(s|y)}ds, \text{ for } y \leq y_0 \tag{1.135}$$

where we have used notations:

$$\tilde{\alpha}(y|y_0) = \int_{y_0}^{y} \alpha(s)ds, \tag{1.136}$$

$$h(s) = g(s) - f'(s). \tag{1.137}$$

(d) The equations of the major hysteresis loop can be derived by taking the limits $x_0 \to \pm\infty$ and $y_0 \to \pm y_{sat}$ in the above equations. Using conditions (1.121) and (1.120) we obtain:

$$x_L^{\infty}(y) = f(y) + \int_{-\infty}^{y} h(s)e^{\tilde{\alpha}(s|y)}ds, \tag{1.138}$$

$$x_U^{\infty}(y) = f(y) - \int_{y}^{\infty} h(s)e^{-\tilde{\alpha}(s|y)}ds. \tag{1.139}$$

It is obvious that $x_U^{\infty}(y) \leq f(y) \leq x_L^{\infty}(y)$. It can also be proven that the two integrals in (1.138) and (1.139) vanish when $y \to \mp\infty$, respectively, which shows that the major hysteresis loop is mainly "shaped" by function f.

(e) It is often assumed in applications that $g(y) = f'(y)$ for all $|y|$ larger than a "closure" value y_{cl}. For $y \geq y_{cl}$ the unloading curve is identical to f; for $y \leq y_{cl}$ the loading curve is identical to f. The difference between the two curves of the major loop is:

$$x_L^{\infty}(y) - x_U^{\infty}(y) = \left[x_L^{\infty}(y_{cl}) - x_U^{\infty}(y_{cl})\right]e^{-\tilde{\alpha}(|y||\text{sgn}(y)y_{cl})}, \text{ for } |y| > y_{cl} \tag{1.140}$$

where $x_L^{\infty}(y_{cl}) - x_U^{\infty}(y_{cl})$ is the difference between the two curves at the "closure" point. Equation (1.140) shows that the difference between the loading and unloading curves is mostly governed by material function α and decreases as $e^{-|y|}$ when $y \to \pm\infty$.

(f) The input at coercivity is given by

$$x_C = \int_{-\infty}^{0} h(s)e^{\tilde{\alpha}(s|y)}ds, \tag{1.141}$$

where we have used (1.119).

(g) The differential susceptibility at coercivity is given by

$$\chi_C = \alpha(0)\int_{-\infty}^{0} h(s)e^{\tilde{\alpha}(s|0)}ds. \tag{1.142}$$

(h) The initial hysteresis curve can be computed by setting $x_0 = y_0 = 0$ in (1.134):

$$x_{ini}(y) = f(y) - f(0)e^{-\tilde{\alpha}(y|0)} - \int_{0}^{y} h(s)e^{\tilde{\alpha}(s|y)}ds, \text{ for } y \geq 0. \tag{1.143}$$

The slope of the initial hysteresis curve at $y = 0$ is given by (1.133) and the curvature by

$$\left.\frac{d^2y}{dx^2}\right|_{x=y=0} = \frac{1}{\alpha(0)h(0)} = \frac{\chi_i}{\alpha(0)[1 - f'(0)\chi_i]}. \tag{1.144}$$

(i) If the input varies periodically between values x_m and x_M it can be shown that the output will eventually very between values y_m and y_M, i.e. the model displays accommodation, which degenerates into a limit cycle (see Fig. 1.23). The conditions that (x_m, x_M) and (y_m, y_M) satisfy when the two points are situated on the limit cycle can be derived by writing Eqs. (1.134) and (1.135) for the ascending and descending curves. We obtain:

$$x_M = f(y_M) + (x_m - f(y_m))e^{-\tilde{\alpha}(y_M|y_m)} + \int_{y_m}^{y_M} h(s)e^{\tilde{\alpha}(s|y_M)}ds, \tag{1.145}$$

Fig. 1.23 Accommodation effects and limit cycle in the Coleman-Hodgdon model

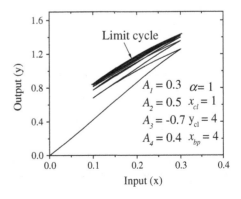

$$x_m = f(y_m) + (x_M - f(y_M))e^{\tilde{\alpha}(y_m|y_M)} - \int_{y_m}^{y_M} h(s)e^{-\tilde{\alpha}(s|y_m)}ds. \tag{1.146}$$

Since $\tilde{\alpha}(y_M|y_m) = -\tilde{\alpha}(y_m|y_M)$, after a few rearrangements we obtain:

$$x_M = f(y_M) + \operatorname{csch}\tilde{\alpha}(y_M|y_m) \int_{y_m}^{y_M} h(s)\sinh\tilde{\alpha}(s|y_m)ds, \tag{1.147}$$

$$x_m = f(y_m) + \operatorname{csch}\tilde{\alpha}(y_M|y_m) \int_{y_m}^{y_M} h(s)\sinh\tilde{\alpha}(s|y_M)ds, \tag{1.148}$$

where $\operatorname{csch}x = (\sinh x)^{-1}$ is the hyperbolic cosecant function. If y_m, and y_M are known than Eqs. (1.147) and (1.148) can be used to compute x_M and x_m respectively. Hence, for any two limits of the output of the limit cycle there is a unique set of inputs x_m and x_M which lead to that limit cycle. The opposite is also true: for any two values of the input x_m and x_M there is a unique limit cycle given by values y_m, and y_M. The proof of the last statement is a bit more difficult since it involves the uniqueness of the solution of system (1.145) and (1.146) and will not be presented here.

(j) Function f is the inverse of the anhysteretic curve. This can be proven by taking the limit $y_M \to y_m$ in (1.148). Indeed, if we consider $y_M = y_m + \varepsilon$, where ε is an infinitesimally small parameter, Eq. (1.148) implies:

$$x_M = f(y_M) - \frac{\int_{y_M-\varepsilon\varepsilon\to 0}^{y_M} h(s)\sinh\tilde{\alpha}(s|y_M - \varepsilon)ds}{\sinh\tilde{\alpha}(y_M - \varepsilon|y_M)}. \tag{1.149}$$

Using l'Hôpital's rule, the limit of the last term in (1.149) is zero when $\varepsilon \to 0$, and we obtain $x_M = f(y_M)$.

(k) The energy lost during one limit cycle defined by points (x_m, y_m) and (x_M, y_M) can be computed as

$$w = \int_{y_m}^{y_M} [x_L(y|x_m, y_m) - x_U(y|x_M, y_M)]dy, \tag{1.150}$$

where $x_L(y)$ and $x_U(y)$ are given by (1.134) and (1.135). To evaluate the integral in (1.2) it is convenient to change the variable $\psi = y - \bar{y}$ where

$$\bar{y} = \frac{y_m + y_M}{2} \tag{1.151}$$

and introduce notations:

$$a_+ = \operatorname{csch}(2\alpha_0\,\delta) \int_{-\delta}^{\delta} h(\psi + y)\sinh\alpha_0(\psi + \delta)d\psi, \tag{1.152}$$

$$a_- = \operatorname{csch}(2\alpha_0\,\delta) \int_{-\delta}^{\delta} h(\psi + y)\sinh\alpha_0(\psi - \delta)d\psi. \tag{1.153}$$

With these notations (1.150) becomes

$$w = \int_{-\delta}^{\delta} \left[a_+ e^{-\alpha_0(\psi+\delta)} - a_- e^{\alpha_0(\psi-\delta)} \right] d\psi$$

$$+ \int_{-\delta}^{\delta} \left[\int_{-\delta}^{\psi} h(s+\bar{y}) \, e^{\alpha_0(s-\psi)} \, ds + \int_{\psi}^{\delta} h(s+\bar{y}) \, e^{\alpha_0(\psi-s)} \, ds \right] d\psi.$$

Performing the integral in the first term and integrating by parts the last two terms one obtains:

$$w = \frac{a_+ - a_-}{\alpha_0} \left(1 - e^{-2\alpha_0\delta}\right) - \frac{e^{-\alpha_0\delta}}{\alpha_0} \int_{-\delta}^{\delta} h(s+\bar{y}) e^{\alpha_0 s} ds$$

$$- \frac{e^{-\alpha_0\delta}}{\alpha_0} \int_{-\delta}^{\delta} h(s+\bar{y}) e^{-\alpha_0 s} ds + \frac{2}{\alpha_0} \int_{-\delta}^{\delta} h(\psi+\bar{y}) d\psi. \qquad (1.154)$$

Since

$$a_+ - a_- = \operatorname{csch}(2\alpha_0\,\delta) \int_{-\delta}^{\delta} h(\psi+y)[\sinh\alpha_0(\psi+\delta) - \sinh\alpha_0(\psi-\delta)]d\psi$$

$$= 2\sinh(\alpha_0\delta)\operatorname{csch}(2\alpha_0\,\delta) \int_{-\delta}^{\delta} h(\psi+\bar{y})\cosh(\alpha_0\psi)d\psi$$

$$= \frac{1}{\cosh(\alpha_0\delta)} \int_{-\delta}^{\delta} h(\psi+\bar{y})\cosh(\alpha_0\psi)d\psi$$

Equation (1.154) becomes

$$w = \frac{2\tanh(\alpha_0\delta)}{\alpha_0 \exp(\alpha_0\,\delta)} \int_{-\delta}^{\delta} h(\psi+\bar{y})\cosh(\alpha_0\psi)d\psi$$

$$+ \frac{2}{\alpha_0} \int_{-\delta}^{\delta} h(\psi+\bar{y})\left[1 - \frac{\cosh(\alpha_0 s)}{\exp(\alpha_0\delta)}\right] d\psi$$

$$= \frac{2}{\alpha_0} \int_{-\delta}^{\delta} h(\psi+\bar{y})\left\{1 - \frac{\cosh(\alpha_0\psi)}{\exp(\alpha_0\delta)}[1 + \tanh(\alpha_0\,\delta)]\right\} d\psi \qquad (1.155)$$

$$= \frac{2}{\alpha_0} \int_{-\delta}^{\delta} h(\psi+\bar{y})\left[1 - \frac{\cosh(\alpha_0\psi)}{\cosh(\alpha_0\delta)}\right] d\psi$$

If the limit cycle is the major hysteresis loop, $\bar{y} = 0$ and $\delta = \infty$, and the energy lost is:

$$w = \frac{4}{\alpha_0} \int_{-\infty}^{\infty} h(s)ds = \frac{4}{\alpha_0} \int_{-\infty}^{\infty} [g(s) - f'(s)]ds. \tag{1.156}$$

It is obvious from the last equation that parameter α_0 can be directly related to the energy lost during one cycle. Also, the sign of $g(s) - f'(s)$ dictates if the system is consuming or generating energy.

(l) The coefficient of the quadratic term that enters into the law of Lord Rayleigh (1.4) can be expresses as a function of the model's parameters as (see (1.144)):

$$b = \frac{1}{\alpha(0)[g(0) - f'(0)]}. \tag{1.157}$$

1.6.3 Parameter Identification Methods

In this section we present a few parameter identification methods for the Coleman-Hodgdon model of hysteresis. To simplify the presentation we assume that function $\alpha(y) = \alpha_0$ is constant. The first method that we present aims at finding functions f and g and parameter α_0 without assuming any particular functional dependence for these functions. The second and third identification methods assume particular functional dependences for f and g and the identification methods aim at finding the constitutive parameters of these functions.

1.6.3.1 Determination of Functions f and g and Parameter α_0

Function $f(y)$ can be identified as the inverse of the anhysteretic curve of the system. Hence, to measure it in practical applications we produce a series of alternating inputs with decreasing magnitude and centered around x_0, and measure the final value of the output y_0. The dependence $x_0(y_0)$ defines function f.

Parameter α_0 can be identified from the initial susceptibility χ_i and the curvature of the initial curve $\left.\frac{d^2y}{dx^2}\right|_{x=y=0}$ by using:

$$\alpha_0 = \frac{\chi_i}{[1 - f'(0)\chi_i]} \left.\frac{d^2x}{dy^2}\right|_{x=y=0}, \tag{1.158}$$

which can be derived from (1.144).

Finally, function $g(y)$ can be identified by using the measured major hysteresis loop. For instance, if we know the loading curve $x_L^\infty(y)$ we can first solve the following Volterra equation of the first kind and compute $h(y)$:

$$x_L^\infty(y) = f(y) + \int_{-\infty}^y h(s)e^{\alpha(s-y)}ds. \tag{1.159}$$

Then, we evaluate

$$g(y) = h(y) + f'(y). \tag{1.160}$$

This identification method has the advantage that it provides the exact form of material functions without pre-assuming any functional dependence.

1.6.3.2 Determination of Parameters α_0, D_1, D_2, and D_3

If we assume that the material functions are given by expressions (1.128) and (1.129) the parameter identification problem can be stated as to find the optimum values of α_0, D_1, D_2, and D_3 that fit the measured hysteresis curves (we assume that y_{bp} and y_{cl} are evaluated from the "closure" point on the major loop, or are set to conveniently large values). Parameter D_1 can be determined as the susceptibility for $|y| > y_{bp}$, which can be measured experimentally. Parameters D_2, D_3, and α_0 can be computed from the initial susceptibility χ_i, coercive input x_C, and remanent output y_R by solving numerically the following system of equations:

$$\chi_i = D_2(1 + D_3), \tag{1.161}$$

$$x_C = \frac{D_2 D_3}{\alpha_0}(1 - e^{-\alpha_0 y_{cl}}), \tag{1.162}$$

$$y_R = \frac{D_3}{\alpha_0}\left[1 - e^{\alpha_0(y_{cl} - y_R)}\right]. \tag{1.163}$$

An alternative way, which avoids solving the above system numerically is to compute α_0 from the initial susceptibility χ_i and the curvature of the initial curve by using (1.158). Then, D_3 is computed from (1.163) and D_2 from either (1.161) or (1.162).

1.6.3.3 Determination of Parameters μ_s, α_0, A_1, A_2, A_3 and A_4

If the material functions are given by expressions (1.131) and (1.132) the parameter identification problem becomes to find the optimum values of μ_s, α_0, A_1, A_2, A_3, and A_4 that fit the measured hysteresis curves (we assume again that y_{bp} and y_{cl} are evaluated from the "closure" point on the major loop or are set to conveniently large values). Parameter μ_s can be determined as the differential susceptibility at saturation ($|y| > y_{cl}$). Parameter α_0 can be identified from the initial susceptibility χ_i and the curvature of the initial curve by using (1.158).

Parameters A_1 and A_2 can be determined by using the susceptibility μ_{cl} at point (x_{cl}, y_{cl}) on the unloading curve of the major loop. We can write $x_{cl} = f(y_{cl}) =$

$A_1 \tan(A_2 y_{cl})$ and $\chi_{cl} = \frac{1}{f'(y_{cl})} = \frac{\cos^2(A_2 y_{cl})}{A_1 A_2}$ which leads to the following transcendental equation for A_2:

$$2A_2 x_{cl} \chi_{cl} = \sin(2A_2 y_{cl}). \tag{1.164}$$

Then, A_1 can be computed by using

$$A_1 = \frac{x_{cl}}{\tan(A_2 y_{cl})}. \tag{1.165}$$

By imposing the condition that $f'(y)$ is continuous at $y = y_{bp}$ we obtain:

$$y_{bp} = \frac{\arccos(A_1 A_2 \chi_{cl})}{A_2}. \tag{1.166}$$

Finally, parameters A_3 and A_4 can be determined using the values of the coercive input and remanent output. Indeed, using (1.141) and the equation of the major loop one can derive:

$$\phi \int_{y_R}^{y_{cl}} \exp\left(-\frac{A_4 s}{y_{cl} - s} - \alpha_0 s\right) ds = \int_0^{y_{cl}} \exp\left(-\frac{A_4 s}{y_{cl} - s} - \alpha_0 s\right) ds, \tag{1.167}$$

$$\frac{1}{A_3} = -\frac{1}{x_C} \int_0^{y_{cl}} \exp\left(-\frac{A_4 s}{y_{cl} - s} - \alpha_0 s\right) ds, \tag{1.168}$$

where $\phi = \frac{x_C e^{\alpha_0 y_R}}{A_1 \tan(A_2 y_R)}$. Equation (1.167) should be solved for A_4 and (1.168) should be used to compute A_3.

An alternative method to compute parameters A_3 and A_4 is to compute A_3 from the initial susceptibility

$$A_3 = 1 - \frac{1}{A_1 A_2 \chi_i} \tag{1.169}$$

and A_4 using one of the Eqs. (1.167) or (1.168).

1.7 Dynamic Models of Hysteresis

Dynamic (or rate-dependent) models of hysteresis take into consideration the rate at which the input variable is changing in time when describing the hysteresis phenomena. The models that are commonly used in the literature for the simulation of dynamic hysteresis are based either on the feedback theory or on the relaxation time approximation. Both types of dynamic models are implemented in HysterSoft©.

1.7.1 Models Based on the Feedback Theory

Dynamic effects can be relatively easily added to any static model of hysteresis by adding a rate-dependent feedback to the system (i.e. using an effective input). For instance, a common approach is to modify the input variable x to depend on the output variable y and on the rate of variation of the output variable y. Denoting the effective value of the input by x_{eff} we have

$$y = \hat{\Gamma} x_{eff}, \tag{1.170}$$

$$x_{eff} = x + F(y, \dot{y}), \tag{1.171}$$

where F is a function of the output variable y and of its derivative with respect to time, $\dot{y}$. Equations (1.170) and (1.171) represent a system of two equations that should be solved for y. In the framework of the Preisach model, this system is a system of integro-differential equations, while in the framework of the Jiles, energetic and Hodgdon models it becomes a system of differential equations.

Although (1.170) and (1.171) can be solved iteratively, the iterations are particularly unstable and they will often diverge even when the norm of function F is much smaller than the norm of x. Therefore, it is recommended to transform (1.170) and (1.171) into an ordinary differential equation (ODE). If we take the derivatives of these equations with respect to time and use the chair rule we obtain

$$\dot{y} = \chi(x_{eff}) \dot{x}_{eff}, \tag{1.172}$$

$$\dot{x}_{eff} = \dot{x} + F_y(y, \dot{y}) \dot{y} + F_{\dot{y}}(y, \dot{y}) \ddot{y}, \tag{1.173}$$

where χ is the differential susceptibility operator, F_y and $F_{\dot{y}}$ are the partial derivatives of F with respect to $\dot{y}$ and $\ddot{y}$. The last two equations can be solved for $\ddot{y}$

$$\ddot{y} = \frac{\frac{\dot{y}}{\chi[x + F(y,\dot{y})]} - \dot{x} - F_y(y, \dot{y}) \dot{y}}{F_{\dot{y}}(y, \dot{y})}. \tag{1.174}$$

HysterSoft© converts this second order ODE to a system of first order ODEs and solves it for any arbitrary function $F(y, \dot{y})$ that can be defined by the user.

1.7.2 Models Based on the Relaxation Time Approximation

In the relaxation time approximation, the output variable $y(t)$ is described by the following first-order differential equation:

$$\frac{dy}{dt} = -\frac{y(t) - \hat{\Gamma} x(t)}{\tau}, \tag{1.175}$$

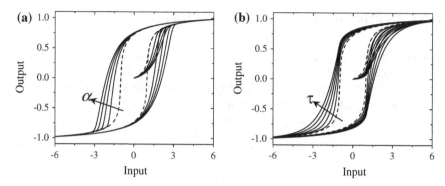

Fig. 1.24 Dynamic hysteresis loops obtained using the energetic model coupled with the feedback theory (**a**) and the relaxation time approximation (**b**). In (**a**) function $F(y, \dot{y}) = -\alpha \dot{y}$, where α increases from 0 to 0.1 in steps of 0.02 in the direction indicated by the arrow; in (**b**) the relaxation time increases from 0 to 0.01 in steps of 0.002 in the direction indicated by the arrow

where τ is a relaxation time parameter and $\hat{\Gamma}$ is any regular static (not rate-dependent) hysteresis operator. The susceptibility can be computed using:

$$\chi(t) = -\frac{y(t) - \hat{\Gamma}x(t)}{\tau \dot{x}(t)}, \qquad (1.176)$$

where $\dot{x}(t)$ is the derivative of the input variable with respect to time. HysterSoft© integrates Eq. (1.176) numerically to compute the output variable y as a function of time.

Figure 1.24 presents rate-dependent simulations obtained using the feedback theory and relaxation time approximation by the energetic model. The static hysteresis loop (obtained at very low rates of variation of the input) is represented by dash line. Notice that the rate-dependent processes increase the width the hysteresis loop and, in this way, the energy lost per cycle.

1.8 Vector Models of Hysteresis

In this section we discuss about vector models of hysteresis in which the input and output variables are two or three-dimensional vectors. The vector models are defined by using superposition of scalar models. This approach was initially introduced by Mayergoyz [77–79] in the framework of the Preisach model and was later generalized to any scalar model of hysteresis [80].

1.8.1 Two-Dimensional Models of Hysteresis

Two-dimensional vector models of hysteresis are constructed as a two-dimensional superposition of scalar models of hysteresis:

$$y(t) = \int_{-\pi/2}^{\pi/2} \hat{\varphi}\, \hat{\Gamma}_\varphi [x(t) \cdot \hat{\varphi}] d\varphi, \qquad (1.177)$$

where $\hat{\varphi}$ is the unit vector along the direction specified by polar angle φ, x is the vector input variable, and $\hat{\Gamma}_\varphi$ is the scalar hysteresis operator along direction φ. $\hat{\Gamma}_\varphi$ can be any scalar hysteresis operator such as the Preisach, Jiles-Atherton, energetic, or other scalar models.

To understand how to implement Eq. (1.177) in numerical simulations, we present the main equations for the vector Jiles-Atherton model of hysteresis. The same analysis can be extended to any other model. In the case of the vector Jiles-Atherton model each model parameter in Eq. (1.47) as well as the anhysteretic function depend on an additional angular parameter φ:

$$\frac{dy}{dx} = \frac{(1 - c_\varphi)\, \delta_\varphi \left[L_\varphi(x + \alpha_\varphi y) - y \right]}{k_\varphi (1 - c_\varphi)\, \mathrm{sgn}(\dot{x}) - \alpha_\varphi \left[L_\varphi(x + \alpha_\varphi y) - y \right]} + c_\varphi \frac{dL_\varphi(x + \alpha_\varphi y)}{dx} \qquad (1.178)$$

where:

$$L_\varphi(x) = y_{sat}\, L\left(\frac{x}{a_\varphi}\right) = y_{sat} \left[\coth\left(\frac{x}{a_\varphi}\right) - \frac{a_\varphi}{x} \right] \qquad (1.179)$$

and δ_φ is equal to 0 if $\dot{x}\left[L_\varphi(x + \alpha_\varphi y) - y \right] \leq 0$ and one otherwise. In the above equations, y_{sat} is the saturation value of the output and a_φ, c_φ, α_φ, and k_φ are functions of φ that can be identified by fitting simulations to experimental results. Notice that in the case of the scalar Jiles-Atherton model of hysteresis a, c, α, and k are the classical model's parameters, however, in the framework of the vector model these parameters become functions of angle φ ranging from $-\pi/2$ to $\pi/2$.

HysterSoft© implements Eq. (1.177) for most scalar hysteresis models presented in this chapter. To simplify the parameter definition along each direction the user can define only two sets of parameters, one set along the x-axis, $a_i(0)$ and one set along the y-axis, $a_i(\pi/2)$, and HysterSoft© will find the values of the parameters along any other axis using the following sinusoidal interpolation

$$a_i(\varphi) = a_i(0) + [a_i(\pi/2) - a_i(0)] \sin\varphi \qquad (1.180)$$

for $-\pi/2 \leq \varphi \leq \pi/2$. An implementation of (1.177) based on the superposition of stop operators can also be found in [14].

1.8.2 Three-Dimensional Models of Hysteresis

The three-dimensional vector model is defined as:

$$y(t) = \oiint_{|r|=1} r\Gamma_{\theta,\phi}[r \cdot x(t)]dS, \qquad (1.181)$$

where the integration is taken over the unit sphere of radius r, and θ and ϕ are the azimuth and inclination angles of vector r. It should be mentioned that model (1.181) was initially introduced in [1] for the Preisach model. Due to the high computational cost required to integrate (1.181), this equation is relatively difficult to implement in real-time simulations for the Preisach model, however, it becomes more manageable when used with other models such as the energetic of Jiles-Atherton models. Given the input vector $x(t) = [x_x(t), \ x_y(t), \ x_z(t)]$, the x, y, and z components of the output can be computed by using the following equations:

$$y_x(t) = \oiint_{|r|=1} \Gamma_{\alpha(\theta,\phi)}[x(t)] \sin^2\theta \cos\phi \; d\theta \; d\phi, \qquad (1.182)$$

$$y_y(t) = \oiint_{|r|=1} \Gamma_{\alpha(\theta,\phi)}[x(t)] \sin^2\theta \sin\phi \; d\theta \; d\phi, \qquad (1.183)$$

$$y_z(t) = \oiint_{|r|=1} \Gamma_{\alpha(\theta,\phi)}[x(t)] \sin 2\theta \; d\theta \; d\phi, \qquad (1.184)$$

where $x(t) = x_x(t) \sin\theta \cos\phi + x_y(t) \sin\theta \sin\phi + x_z(t) \cos\theta$.

In the case of three-dimensional vector models of hysteresis the parameters of the scalar models depend on two angles φ and θ. For instance, in the case of the Jiles-Atherton model:

$$\frac{dy}{dx} = \frac{(1 - c_{\varphi,\theta}) \; \delta_{\varphi,\theta}[L_{\varphi,\theta}(x + \alpha_{\varphi,\theta}y) - y]}{k_{\varphi,\theta}(1 - c_{\varphi,\theta}) \; \text{sgn}(\dot{x}) - \alpha_{\varphi,\theta}[L_{\varphi,\theta}(x + \alpha_{\varphi,\theta}y) - y]} + c_{\varphi,\theta}\frac{dL_{\varphi,\theta}(x + \alpha_{\varphi,\theta}y)}{dx}$$

$$(1.185)$$

and the Langevin function becomes:

$$L_{\varphi,\theta}(x) = M_s \; L\left(\frac{x}{a_{\varphi,\theta}}\right) = y_{sat}\left[\coth\left(\frac{x}{a_{\varphi,\theta}}\right) - \frac{a_{\varphi,\theta}}{x}\right]. \qquad (1.186)$$

1.9 Summary

This chapter offered an overview of the hysteresis models that are used throughout the book. After a short general classification of hysteresis models and parameter identification methods, the rectangular hysteresis operator was introduced.

Then, the chapter focused on summarizing the main equations, properties, and characteristics of the Preisach, energetic, Jiles-Atherton, Coleman-Hodgdon, and Bouc-Wen models. Particular attention was given to the analytical description of the general properties of hysteresis curves such as differential susceptibilities, remanence, coercivity, saturation, anhysteretic curve, energy lost, stability, accommodation, and limit cycle for each model. The second part of the chapter presented two techniques for the modeling of rate-dependent hysteresis, one based on the feedback (effective field) theory and the other one on the relaxation time approximation. Finally, a unified theory of vector models was presented; this theory can be applied to generalize any scalar model of hysteresis to describe vector systems. The presentation was mostly addressed to applied scientists and engineers and we recommend the readers interested in a more precise mathematical formulation of hysteresis phenomena to study monographs [81–85].

References

1. Mayergoyz, I. (1991). *Mathematical models of hysteresis*. New York: Springer-Verlag.
2. Bouc, R. (1971). Mathematical model for hysteresis. *Acustica, 24*, 16–25.
3. Wen, Y. K. (1976). Method for random vibration of hysteretic systems. *Journal of the Engineering Mechanics Division-ASCE, 102*, 249–263.
4. Duhem, P. (1897). Die dauernden Aenderungen und die Thermodynamik. *Zeitschrift für Physikalische Chemie, 22*, 543.
5. Jiles, D. C., & Atherton, D. L. (1983). Ferromagnetic hysteresis. *IEEE Transactions on Magnetics, 19*, 2183–2185.
6. Jiles, D. C., & Atherton, D. L. (1984). Theory of ferromagnetic hysteresis. *Journal of Applied Physics, 55*, 2115–2120.
7. SPICE. (University of California at Berkeley), *The SPICE Webpage*. Retrieved from http://bwrc.eecs.berkeley.edu/classes/icbook/spice/.
8. Coleman, B. D., & Hodgdon, M. L. (1987). On a class of constitutive relations for ferromagnetic hysteresis. *Archive for Rational Mechanics and Analysis, 99*, 375–396.
9. Coleman, B. D., & Hodgdon, M. L. (1986). A constitutive relation for rate-independent hysteresis in ferromagnetically soft materials. *International Journal of Engineering Science, 24*, 897–919.
10. Hauser, H. (1994). Energetic model of ferromagnetic hysteresis. *Journal of Applied Physics, 75*, 2584–2596.
11. Harrison, R. G. (2009). Physical theory of ferromagnetic first-order return curves. *IEEE Transactions on Magnetics, 45*, 1922–1939.
12. Harrison, R. G. (2011). Positive-feedback theory of hysteretic recoil loops in hard ferromagnetic materials. *IEEE Transactions on Magnetics, 47*, 175–191.
13. Bergqvist, A., et al. (1997). Experimental testing of an anisotropic vector hysteresis model. *IEEE Transactions on Magnetics, 33*, 4152.
14. Leite, J. V., et al. (2005). A new anisotropic vector hysteresis model based on stop hysterons. *IEEE Transactions on Magnetics, 41*, 1500–1503.
15. Matsuo, T., et al. (2004). Stop model with input-dependent shape function and its identification methods. *IEEE Transactions on Magnetics, 40*, 1776–1783.

16. de Almeida, L. A. L., et al. (2003). Limiting loop proximity hysteresis model. *IEEE Transactions on Magnetics, 39*, 523–528.
17. Takács, J. (2003). *Mathematics of hysteretic phenomena: The T(x) model for the description of hysteresis*. Weinheim: Wiley.
18. Kucuk, I. (2006). Prediction of hysteresis loop in magnetic cores using neural network and genetic algorithm. *Journal of Magnetism and Magnetic Materials, 305*, 423–427.
19. Cao, S. Y., et al. (2006). Modeling dynamic hysteresis for giant magnetostrictive actuator using hybrid genetic algorithm. *IEEE Transactions on Magnetics, 42*, 911–914.
20. Rayleigh, L. (1887). On the behaviour of iron and steel under the operation of feeble magnetic forces. *Philosophical Magazine, 23*, 225–248.
21. Ikhouane, F., & Rodellar, J. (2007). *Systems with hysteresis*. Chichester: John Wiley & Sons, Ltd.
22. Xue, X. M., et al. (2010). Parameter estimation for the phenomenological model of hysteresis using efficient genetic algorithm. *ISCM II and EPMESC XII, 1233*, 713–717.
23. Sun, Q., et al. (2009). Parameter estimation and its sensitivity analysis of the MR damper hysteresis model using a modified genetic algorithm. *Journal of Intelligent Material Systems and Structures, 20*, 2089–2100.
24. Liu, G. J., & Chan, C. H. (2007). Hysteresis identification and compensation using a genetic algorithm with adaptive search space. *Mechatronics, 17*, 391–402.
25. Peng, L., & Wang, W. (2007). Adaptive genetic algorithm with heuristic weighted crossover operator based hysteresis identification and compensation. *2007 IEEE International Conference on Control and Automation* (Vol. 1–7, pp. 3260–3264).
26. Zheng, J. J., et al. (2007). Hybrid genetic algorithms for parameter identification of a hysteresis model of magnetostrictive actuators. *Neurocomputing, 70*, 749–761.
27. Chwastek, K., & Szczyglowski, J. (2006). Identification of a hysteresis model parameters with genetic algorithms. *Mathematics and Computers in Simulation, 71*, 206–211.
28. Zidaric, B., & Miljavec, D. (2005). Nested genetic algorithms in determination of Jiles-Atherton hysteresis model parameters for soft-magnetic composite materials. *Informacije Midem-Journal of Microelectronics Electronic Components and Materials, 35*, 92–96.
29. Cao, S. Y. et al. (2005) Parameter identification of strain hysteresis model for giant magnetostrictive actuators using a hybrid genetic algorithm. *ICEMS 2005: Proceedings of the Eighth International Conference on Electrical Machines and Systems, 1–3*, 2009–2012.
30. Chan, C. H., & Liu, G. J. (2004). Actuator hysteresis identification and compensation using an adaptive search space based genetic algorithm. *Proceedings of the 2004 American Control Conference, 1–6*, 5760–5765.
31. Naghizadeh, R. A., & Vajidi, B. (2011). Parameter identification of Jiles-Atherton model using SFLA. *COMPEL: The International Journal for Computation and Mathematics in Electrical and Electronic Engineering, 31*, 1293–1309.
32. Ye, M. Y., & Wang, X. D. (2009). Parameter identification of hysteresis model with improved particle swarm optimization. *Proceedings of the 21st Chinese Control and Decision Conference, 1–6*, 415–419.
33. Ye, M. Y., & Wang, X. D. (2007). Parameter estimation of the Bouc-Wen hysteresis model using particle swarm optimization. *Smart Materials and Structures, 16*, 2341–2349.
34. Fulginei, F. R., et al. (2007). Symbiotic evolutionary algorithm for the Preisach hysteresis model identification. *International Journal of Applied Electromagnetics and Mechanics, 25*, 681–687.
35. Andrei, P., et al. (2007). Identification techniques for phenomenological models of hysteresis based on the conjugate gradient method. *Journal of Magnetism and Magnetic Materials, 316*, E330–E333.

36. Mayergoyz, I. (2003). *Mathematical models of hysteresis and their applications: Electromagnetism.* San Diego: Academic Press.
37. Pike, C. R., et al. (2005). First-order reversal curve diagram analysis of a perpendicular nickel nanopillar array. *Physical Review B, 71,* 134407.
38. Della Torre, E., & Bennett, L. H. (1998). A Preisach model for aftereffect. *IEEE Transactions on Magnetics, 34,* 1276–1278.
39. Cardelli, E., et al. (2000). Direct and inverse Preisach modeling of soft materials. *IEEE Transactions on Magnetics, 36,* 1267–1271.
40. Fry, R. A., et al. (2000). Preisach modeling of aftereffect in a magneto-optical medium with perpendicular magnetization. *Physica B-Condensed Matter, 275,* 50–54.
41. Patel, U. D., & Della Torre, E. (2001). Fast computation of the inverse CMH model. *Physica B, 306,* 178–184.
42. Reimers, A., et al. (2001). Implementation of the preisach DOK magnetic hysteresis model in a commercial finite element package. *IEEE Transactions on Magnetics, 37,* 3362–3365.
43. Reimers, A., & Della Torre, E. (2002). Implementation of the simplified vector model. *IEEE Transactions on Magnetics, 38,* 837–840.
44. Cardelli, E., et al. (2004). Modeling of laminas of magnetic iron with a reduced vector Preisach model. *Physica B-Condensed Matter, 343,* 171–176.
45. Della Torre, E., et al. (2004). Differential equation model for accommodation magnetization. *IEEE Transactions on Magnetics, 40,* 1499–1505.
46. Burrascano, P., et al. (2006). Vector hysteresis model at micromagnetic scale. *IEEE Transactions on Magnetics, 42,* 3138–3140.
47. Della Torre, E., et al. (2006). Vector modeling—part I: Generalized hysteresis model. *Physica B-Condensed Matter, 372,* 111–114.
48. Della Torre, E., et al. (2006). Vector modeling—part II: Ellipsoidal vector hysteresis model. Numerical application to a 2D case. *Physica B-Condensed Matter, 372,* 115–119.
49. Della Torre, E., & Cardelli, E. (2007). The coordinated vector model. *COMPEL-the International Journal for Computation and Mathematics in Electrical and Electronic Engineering, 26,* 327–333.
50. Della Torre, E., et al. (2008). A model for vector accommodation. *Physica B-Condensed Matter, 403,* 496–499.
51. Cardelli, E., et al. (2009). Analysis of a unit magnetic particle via the DPC model. *IEEE Transactions on Magnetics, 45,* 5192–5195.
52. Stancu, A., et al. (2005). New Preisach model for structured particulate ferromagnetic media. *Journal of Magnetism and Magnetic Materials, 290,* 490–493.
53. Mitchler, P. D., et al. (1996). Henkel plots in a temperature and time dependent Preisach model. *IEEE Transactions on Magnetics, 32,* 3185–3194.
54. Mitchler, P. D., et al. (1999). Interactions and thermal effects in systems of fine particles: A Preisach analysis of CrO_2 audio tape and magnetoferritin. *IEEE Transactions on Magnetics, 35,* 2029–2042.
55. Spinu, L., et al. (2001). Time and temperature-dependent Preisach models. *Physica B, 306,* 166–171.
56. Matsuo, T., et al. (2003). Application of stop and play models to the representation of magnetic characteristics of silicon steel sheet. *IEEE Transactions on Magnetics, 39,* 1361–1364.
57. Matsuo, T., & Shimasaki, M. (2005). Representation theorems for stop and play models with input-dependent shape functions. *IEEE Transactions on Magnetics, 41,* 1548–1551.
58. Bergqvist, A., et al. (1997). Experimental testing of an anisotropic vector hysteresis model. *IEEE Transactions on Magnetics, 33,* 4152–4154.

59. Deane, J. H. B. (1994). Modeling the dynamics of nonlinear inductor circuits. *IEEE Transactions on Magnetics, 30*, 2795–2801.

60. Cadence. (2005), *PSPICE*. Retrieved from http://www.cadence.com/.

61. Jiles, D. C., & Atherton, D. L. (1986). Theory of ferromagnetic hysteresis. *Journal of Magnetism and Magnetic Materials, 61*, 48–60.

62. Sablik, M. J., & Jiles, D. C. (1988). A model for hysteresis in magnetostriction. *Journal of Applied Physics, 64*, 5402–5404.

63. Hauser, H. et al. (Oct 2007). Including effects of microstructure and anisotropy in theoretical models describing hysteresis of ferromagnetic materials. *Applied Physics Letters, 91*, 172512.

64. Jiles, D. C., et al. (1992). Numerical determination of hysteresis parameters for the modeling of magnetic-properties using the theory of ferromagnetic hysteresis. *IEEE Transactions on Magnetics, 28*, 27–35.

65. Hauser, H. (1995). Energetic model of ferromagnetic hysteresis. 2. Magnetization calculations of (110)[001] fesi sheets by statistic domain behavior. *Journal of Applied Physics, 77*, 2625–2633.

66. Hauser, H. (2004). Energetic model of ferromagnetic hysteresis: Isotropic magnetization. *Journal of Applied Physics, 96*, 2753–2767.

67. Andrei, P., & Adedoyin, A. (Apr 2009). Noniterative parameter identification technique for the energetic model of hysteresis. *Journal of Applied Physics, 105*, 07D523.

68. Baber, T. T., & Wen, Y. K. (1981). Random vibration of hysteretic, degrading systems. *Journal of the Engineering Mechanics Division-ASCE, 107*, 1069–1087.

69. Baber, T. T., & Noori, M. N. (1985). Random vibration of degrading, pinching systems. *Journal of Engineering Mechanics-ASCE, 111*, 1010–1026.

70. Baber, T. T., & Noori, M. N. (1986). Modeling general hysteresis behavior and random vibration application. *Journal of Vibration Acoustics Stress and Reliability in Design-Transactions of the ASME, 108*, 411–420.

71. Hodgdon, M. L. (1991). A constitutive relation for hysteresis in superconductors. *Journal of Applied Physics, 69*, 2388–2396.

72. Hodgdon, M. L. (1988). Mathematical theory and calculations of magnetic hysteresis curves. *IEEE Transactions on Magnetics, 24*, 3120–3122.

73. Stancu, A., et al. (1997). Models of hysteresis in magnetic cores. *Journal de Physique IV, 7*, 209–210.

74. Hodgdon, M. L. (1991). Computation of superconductor critical current densities and magnetization curves. *Journal of Applied Physics, 69*, 4904–4906.

75. Boley, C. D., & Hodgdon, M. L. (1989). Model and simulations of hysteresis in magnetic cores. *IEEE Transactions on Magnetics, 25*, 3922–3924.

76. Andrei, P. (1997). Phenomenological models for the study of ferromagnetic and fertic materials (in Romanian). B.S., Physics Department, Alexandru Ioan Cuza University, Iasi.

77. Mayergoyz, I. D., & Friedman, G. (1987). Isotropic vector Preisach model of hysteresis. *Journal of Applied Physics, 61*, 4022–4024.

78. Mayergoyz, I. D., & Friedman, G. (1987). On the integral-equation of the vector Preisach hysteresis model. *IEEE Transactions on Magnetics, 23*, 2638–2640.

79. Mayergoyz, I. D. (1988). Vector Preisach hysteresis models. *Journal of Applied Physics, 63*, 2995–3000.

80. Andrei, P., & Adedoyin, A. (Apr 2008). Phenomenological vector models of hysteresis driven by random fluctuation fields. *Journal of Applied Physics, 103*, 07D913.

81. Krasnosel'skii, M. A., & Pokrovskii, A. (1989). Systems with hysteresis, Nauka.

82. Visintin, A. (1994). Differential models of hysteresis. Berlin: Springer.

83. Brokate, M., & Sprekels, J. (1996). Hysteresis and phase transitions, Springer.
84. Krejčí, P. (1996) Hysteresis, convexity and dissipation in hyperbolic equations. Tokyo: Gakkotosho Co., Ltd.
85. Mayergoyz, I. D., & Bertotti G. (2006). Science of hysteresis, Academic press

Chapter 2
Noise and Stochastic Processes

2.1 Noise

This section is aimed at familiarizing the reader with most common noise models
in hysteretic systems, emphasizing disruptive and constructive effects of noise on
system behavior. It also addresses the main numerical techniques used for noise
simulation.

2.1.1 Introduction

Everybody hates noise while the world tends to become even noisier. There is an
increasing amount of evidence regarding the negative effects of noise on human
health and environment while the measures taken against noise pollution proved to
be inefficient. For example, World Health Organization (WHO) recommends an
average bellow 35 dB for continuous background noise in hospitals but most of the
measurements presented by numerous scientific articles have indicated average
noise levels between 50 and 70 dB featuring generally flat spectra over the
60–2000 Hz band. A relevant survey on this topic is provided by Busch-Vishniac
and his colleagues from Johns Hopkins University in Ref. [1] indicating a trend of
increasing noise level in hospitals over the last half a century in spite of WHO
recommendations and the implementations of modern noise reduction techniques.
While the general public is much more aware of and concerned about this acoustic
noise, scientists and engineers are most commonly challenged by electromagnetic
noise, from the cosmic microwave background radiation generated by *Big Bang* to
the electronic noise generated by all electronic circuits.

One of the first areas that addressed noise problem systematically was communication. It is well-known that transmitted signal can be significantly altered by the
noise existent in a communication channel due to the thermal agitation of molecules,
the interference with other signals moving simultaneously through the same channel
or neighboring ones, defects of the material structure, etc. Various techniques are

M. Dimian and P. Andrei, *Noise-Driven Phenomena in Hysteretic Systems*,
Signals and Communication Technology 218, DOI: 10.1007/978-1-4614-1374-5_2,
© Springer Science+Business Media New York 2014

used to reduce these disruptive effects of noise added to the signal such as filtering the noise out, using redundant coding routines of the transmitted signal, controlling the transmission environment, or additional processing of the received signal [2, 3].

The interest in noise analysis has significantly expanded during the last years with the advancement in nanoscience and nanotechnology. Noise is playing a major role in the behavior of nanoscale systems and its effects are increasingly pronounced with the decrease in system size. Let us consider the case of magnetic recording nanotechnology, where thermal noise poses fundamental limits for further improvements in magnetic data storage density. As predicted theoretically by Néel-Arrhenius theory [4] and proved experimentally by Wernsdorfer and his collaborators [5, 6], the switching fields of magnetic nanoparticles decrease with the increase in the temperature up to some blocking temperature when magnetization becomes completely unstable. For a 3 nm cubo-octahedral Co nanoparticle considered in the experiments, the blocking temperature is about 14 K, and, in general, for nanoparticles with diameters below 20 nm the blocking temperatures were found to be below 200 K [6–8]. It is apparent that this superparamagnetic effect found in magnetic nanoparticles and nanograins limits the advances in magnetic data storage density under the current paradigm. On the other hand, thermal noise may also play a positive role in achieving higher storage densities by using the recently developed technology referred as thermally assisted magnetic recording [9, 10]. While high anisotropy media are used in order to provide sufficiently stable magnetic bits at room temperature, the data are recorded at high temperature which reduces significantly the coercive field to values accessible by the current recording heads (see Fig. 2.1). It is foreseen that this recording nanotechnology will

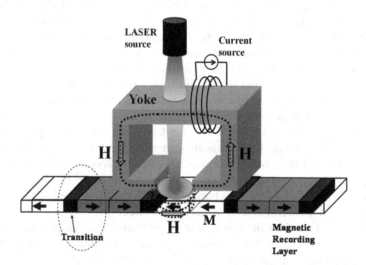

Fig. 2.1 Schematic representation of a heat assisted magnetic recording system. The laser is heating the memory cell in order to generate fast thermal induced switching of the magnetization at magnetic fields accessible to the recording head (formed by a current source and a yoke which amplifies the field in the air-gape)

be the key for exceeding 1 Tb/in^2 storage density. In conclusion, the thermal noise in nanoscale devices might jeopardize the future development of several nano-technologies, such as magnetic data recording, but it could also provide the keys for solving the challenges encountered by such technologies.

Since noise can have only negative effects in linear systems, its potential benefits seem rather counterintuitive and have been overlooked by researchers for a long period of time [11]. However, the recent studies on stochastically driven nonlinear systems proved that such phenomena are quite common and their applications range from signal processing (dithering effect) and nanotechnology (thermal assisted magnetic recording; noise enhanced characteristics of nanotube transistors) to neuroscience (neuron models) and climate models (possible explanations of ice age) [11–15]. These constructive aspects of noise in hysteretic systems will be addressed in Chap. 6.

2.1.2 Wiener Process

Almost two centuries ago, Scottish botanist Robert Brown was the first to sys-tematically analyze the perpetual irregular motion of small pollen grains sus-pended in water. In general, this random drifting, known today as Brownian motion, was observed for any small particles suspended in a fluid. A pertinent explanation of these phenomena did not come until the beginning of twentieth century, when Albert Einstein published his first paper on Brownian motion that contains the key ideas for developing a stochastic analysis. Wiener construction can be seen as the limiting case of the particle Brownian motion as the number of particles and collision rates go to infinity. In addition to its practical applications in the various areas such as physics, biology and finance, Wiener process plays a vital role in stochastic analysis being the foundation for defining more complicated stochastic processes.

The transition probability function of the Wiener process satisfies the following Fokker-Planck equation (FPE) [16]:

$$\frac{\partial}{\partial t} p(x, t | x_0, t_0) = \frac{1}{2} \frac{\partial^2}{\partial x^2} p(x, t | x_0, t_0), \tag{2.1}$$

with the initial condition $p(x, t_0 | x_0, t_0) = \delta(x - x_0)$. By applying Fourier trans-formation with respect to x variable, $\tilde{p}(s, t | x_0, t_0) = \int_{-\infty}^{\infty} p(x, t | x_0, t_0) e^{isx} dx$, Eq. (2.1) becomes:

$$\frac{\partial}{\partial t} \tilde{p}(s, t | x_0, t_0) = -\frac{1}{2} s^2 \tilde{p}(s, t | x_0, t_0), \tag{2.2}$$

subject to initial condition $\tilde{p}(s, t_0 | x_0, t_0) = e^{isx_0}$, which can be simply solved by separation of variables leading to the following solution:

Fig. 2.2 Evolution of the transition probability density for a Wiener process

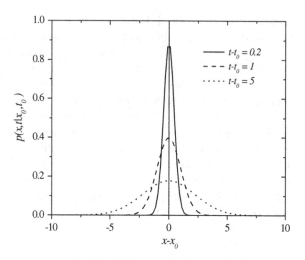

$$\tilde{p}(s, t | x_0, t_0) = e^{isx_0 - \frac{1}{2}s^2(t - t_0)} \tag{2.3}$$

By Fourier inversion, the solution of Eq. (2.1) can be obtained as follows:

$$p(x, t | x_0, t_0) = \frac{1}{\sqrt{2\pi(t - t_0)}} e^{-\frac{(x - x_0)^2}{2(t - t_0)}} \tag{2.4}$$

As a result, the transition probability for the Wiener process has a Gaussian shape with the center in x_0 and variance $(t - t_0)$. Thus the initial δ – distribution is spread in time (see Fig. 2.2) and the variance becomes infinite as $t \to \infty$ indicating a high irregularity of the sample paths, as it is illustrated in Fig. 2.3a.

Although Wiener process has continuous paths, they are almost everywhere not differentiable and have unbounded variation on any finite time interval [17]. If we return to the physical origins of the Wiener process, this indicates an infinite speed

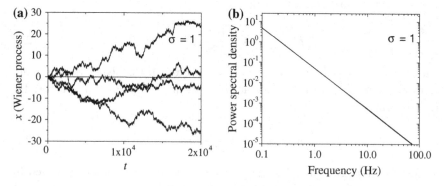

Fig. 2.3 a Simulated sample paths of Wiener process with $\sigma = 1$ starting at $x_0 = 0$; **b** Power spectral density of Wiener process with $\sigma = 1$ starting at $x_0 = 0$

of the Brownian particle, which is obviously one of the drawbacks of Wiener model. A more realistic model, but also more complex, of the Brownian motion is Ornstein-Uhlenbeck process which will be analyzed in the Sect. 2.1.5.

Another important property of Wiener process is the autocorrelation which is defined as follows:

$$\langle X(t_1) \cdot X(t_2)|x_0, t_0\rangle = \iint\limits_{R^2} x_1 x_2 p(x_1, t_1; x_2, t_2 | x_0, t_0) dx_1 dx_2 \qquad (2.5)$$

By using the Markovian property of Wiener process and assuming that $t_2 > t_1$, the autocorrelation function can be written as follows:

$$\langle X(t_2) \cdot X(t_1)|x_0, t_0\rangle = \iint\limits_{R^2} x_1 p(x_2, t_2 | x_1, t_1) p(x_1, t_1 | x_0, t_0) dx_1 dx_2 \qquad (2.6)$$

By taking into account the expression for the first two moments of transition probability density (2.4), one can simply derive the following:

$$\langle X(t_2) \cdot X(t_1)|x_0, t_0\rangle = \int_R \langle X(t_2)|x_1, t_1\rangle x_1 p(x_1, t_1 | x_0, t_0) dx_1$$
$$= \frac{1}{\sqrt{2\pi(t_1 - t_0)}} \int_R x_1^2 e^{-\frac{(x_1 - x_0)^2}{2(t_1 - t_0)}} dx_1 \qquad (2.7)$$

As a result, the Wiener autocorrelation function is

$$\langle X(t_2) \cdot X(t_1)|x_0, t_0\rangle = (t_1 - t_0) + x_0^2 \qquad (2.8)$$

When t_2 is smaller than t_1, the Wiener autocorrelation is obtained by simply replacing t_1 with t_2 in the final formula. Thus, the general expression can be written as:

$$\langle X(t_2) \cdot X(t_1)|x_0, t_0\rangle = (\min\{t_1, t_2\} - t_0) + x_0^2 \qquad (2.9)$$

It is apparent from formula (2.9) that the Wiener process is not stationary, and consequently the power spectral density cannot be expressed in the classical terms as Fourier transform of autocorrelation function. However, a time-dependent spectrum can be defined according to the Wigner-Ville approach:

$$S_{WV}(t, \omega) = \int_{-\infty}^{\infty} x(t + \tau/2) x^*(t - \tau/2) e^{-i\tau\omega} d\tau \qquad (2.10)$$

where the equality is understood in the mean-square sense, x^* denotes the complex conjugate of x, and $i = \sqrt{-1}$. For real valued processes only the real part of the formula is considered. Applying this formula to the Wiener process with $t_0 = 0$ and $x_0 = 0$ it is found that:

$$S_{WV}(t, \omega) = 2\left(\frac{\sin -\omega t-}{\omega}\right)^2 u(t) \tag{2.11}$$

where $u(t)$ is the step function simply pointing out that $t > 0$.

It is also customary to define the average spectrum over the certain interval of length T:

$$S_{WV}(\omega) = \frac{1}{T}\int_0^T S_{WV}(t, \omega)dt \tag{2.12}$$

In the case of the Wiener process, the average spectrum is inverse proportional to ω^2 as suggested by the simulation presented in Fig. 2.3b.

By using the autocorrelation formula (2.9) and simple algebraic calculations it can be proven that increments of the Wiener process, $X(t)-X(s)$, are uncorrelated and have variance $(t-s)$. Since the difference of two Gaussian variables is also Gaussian, we can conclude that the increments of Wiener process are independent and identically distributed (i.i.d.) Gaussian random variables with zero mean and variance $(t-s)$. In addition to the relation with white noise and stochastic differential equations, this property is also useful for the numerical simulation of the Wiener process. Thus a random number Z is generated at each time step according to a standardized normal distribution $N(0, 1)$ and the increments of the sample paths are computed according to the formula

$$x(t_n) - x(t_{n-1}) = Z\sqrt{(t_n - t_{n-1})} \tag{2.13}$$

Simulations of the sample paths using this procedure are presented in Fig. 2.3a.

2.1.3 Itô Stochastic Integral and Differential Equations

Stochastic calculus aimed at extending the benefits of deterministic calculus to the area of stochastic processes. After several less successful approaches developed by Wiener and his collaborators, the Japanese mathematician Kyosi Itô introduced a kind of Riemann-Stieljes integral having Wiener process as integrand and proved the convergence of the integral sums. For the introduction of Itô's construction let us denote Wiener process by $W(t)$ and consider a left-continuous function of time denoted by $G(t)$, which can be either deterministic or stochastic. The stochastic integral $\int_{t_0}^t G(t')dW(t')$ is defined by using Riemann-Stieljes approach as limit of the integral sums:

$$S_n = \sum_{i=1}^n G(t_{i-1})[W(t_i) - W(t_{i-1})] \tag{2.14}$$

over all possible partitions $(t_0 \leq t_1 \leq t_2 \leq \cdots \leq t_{n-1} \leq t_n = t)$ of the interval $[t_0, t]$, with n approaching infinity. The limit is considered in the mean square sense over the probability space Ω, i.e. $\lim\limits_{n\to\infty} \int_\Omega [S_n(\omega) - S(\omega)]^2 \, p(\omega)d\omega = 0$. The convergence of Itô's integral sums is rather counterintuitive knowing that $W(t)$ is almost nowhere differentiable and have unbounded variation on any finite time interval. However, let us note that the choice of the intermediate points is restricted to be the left limits of the partition intervals which is essential in obtaining the convergence of the stochastic integral sums.

The construction of a stochastic integral opens the way towards defining and characterizing more complex stochastic processes via stochastic differential equations. Thus, a stochastic process $X(t)$ is considered a solution of Itô's stochastic differential equation (SDE) written as:

$$dX(t) = b[X(t), t]dt + \sigma[X(t), t]dW(t) \tag{2.15}$$

if for all t and t_0,

$$X(t) = X(t_0) + \int_{t_0}^t b[X(t'), t']dt' + \int_{t_0}^t \sigma[X(t'), t']dW(t') \tag{2.16}$$

where b is the drift coefficient and σ is the diffusion coefficient.

The existence and uniqueness of the solution for this equation in a time interval $[t_0, T]$ subject to a given initial condition can be proven [18] under the following restrictions imposed on the equation coefficients:

- *Lipschitz condition*: a K_L exists such that for all x and y, and all t in the interval $[t_0, T]$,

$$|b(x, t) - b(y, t)| + |\sigma(x, t) - \sigma(y, t)| \leq K_L |x - y|; \tag{2.17}$$

- *Growth condition* : a K_G exists such that for all x, and for all t in the interval $[t_0, T]$,

$$|b(x, t)|^2 + |\sigma(x, t)|^2 \leq K_G \left(1 + |x|^2\right). \tag{2.18}$$

The Lipschitz condition is usually satisfied by the stochastic differential equation used in practice, but the growth conditions is often violated. This does not preclude the existence of a solution rather it indicates the solution is unbounded on the given finite time interval.

In order to connect the two approaches introduced in this chapter to describe a stochastic process, let us mention that the time evolution of the probability density characterizing the stochastic process defined by (2.15) is the solution of FPE:

$$\frac{\partial}{\partial t} p(x, t|x_0, t_0) = -\frac{\partial}{\partial x}[b(x, t)p(x, t|x_0, t_0)] + \frac{1}{2}\frac{\partial^2}{\partial x^2}\left[\sigma^2(x, t)p(x, t|x_0, t_0)\right] \tag{2.19}$$

subject to a δ-initial condition $p(x, t_0|x_0, t_0) = \delta(x - x_0)$ and given boundary conditions.

The generalization of SDE and FPE to multi-dimensional stochastic processes $X(t)$ is quite straightforward. Thus, multi-dimensional Itô's SDE reads:

$$dX(t) = b[X(t), t]dt + \sigma[X(t), t]dW(t) \qquad (2.20)$$

where b is the drift vector function and σ is the diffusion tensor function, while $W(t)$ is the standard multi-dimensional Wiener process. The associated FPE is:

$$\frac{\partial}{\partial t} p(x, t|x_0, t_0) = -\sum_{i=1}^{n} \frac{\partial}{\partial x_i} [b_i(x, t) p(x, t|x_0, t_0)]$$

$$+ \frac{1}{2} \sum_{i,j=1}^{n} \frac{\partial^2}{\partial x_i \partial x_j} \left[\left(\sum_{k=1}^{n} \sigma_{ik}(x, t) \sigma_{kj}(x, t) \right) p(x, t|x_0, t_0) \right] \qquad (2.21)$$

where b_i and σ_{ij} are elements of the drift vector and the diffusion tensor, respectively.

2.1.4 White Noise

White noise is a stochastic process formed by uncorrelated random variables with constant mean and nonzero variance. It is apparent that the autocorrelation of a white noise is a delta function and consequently, its power spectral density is constant. This explains its name drawn from "white light" which has a flat power spectral density over the visible electromagnetic frequency band.

The definition of white noise places no restriction on the probability distribution functions describing the random variables, except the constant mean and variance. Usually, the notion of white noise is used in a stronger form when the component random variables are i.i.d. The numerical implementation of white noise used in this book is based on this idea but various probability density functions (p.d.f.) are considered. Sample of white noise simulations obtained for Gaussian, uniform, Cauchy, and Laplace distributions are shown in Fig. 2.4. Although there is an infinite variety of white noises, the Gaussian type is the overwhelming common noise model in science and engineering, so common that people use it by default when refering to white noise. That is partially related to the central limit theorem of the probability theory stating that the average of a large number of independent random variables converges, under some conditions, to a random variable with Gaussian distribution [19]. In addition, white Gaussian noise (WGN) is the formal derivative of the Wiener process, so it plays a central role in the theory of stochastic differential equation, as it is next discussed.

Let us now recall a definition of the generalized derivative for a deterministic function. The generalized derivative of a function w integrable over a real domain D exists and is denoted by dw/dt if the following equality is satisfied for all infinitely differentiable functions g with compact support in D:

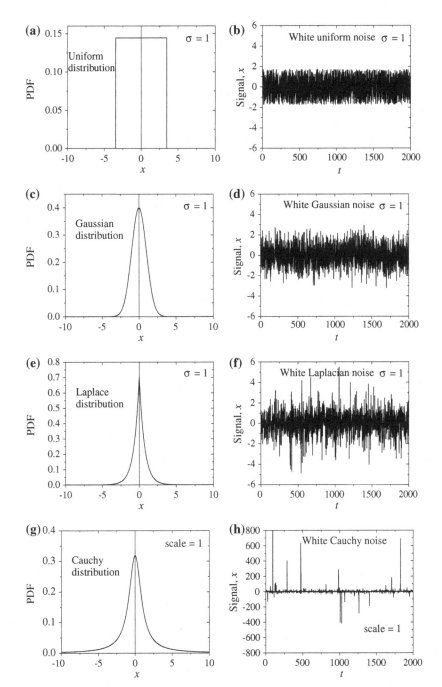

Fig. 2.4 Sample paths of white noise for the uniform, Gaussian, Laplace, and Cauchy distributions, which are represented on the *left hand side*

$$\int_D g(t)\frac{dw}{dt}(t)dt = -\int_D w(t)\frac{dg}{dt}(t)dt \qquad (2.22)$$

For differentiable functions the above formula is nothing else than the integration by parts, so the classical derivative is equal (almost everywhere) to the generalized derivative. It is known that the Wiener process has continuous paths but they are almost everywhere not differentiable in the classical sense. Nevertheless, the generalized derivatives exist and they are expected to be realizations of a WGN since the derivative should involve increments of the Wiener process which are known to be independent and Gaussian. As a result, the stochastic differential Eq. (2.15) is often written in the following form, known as Langevin's equation:

$$\frac{dX}{dt}(t) = b[X(t), t] + \sigma[X(t), t]\xi(t) \qquad (2.23)$$

where $\xi(t)$ is a WGN.

In the end of this section, let us mention that white noise bears a physical inconsistency, namely it requires infinite energy. It is obvious that integrating the constant power spectral density over an infinite frequency band would result an infinite quantity. In practice, a random signal is considered "white noise" if it has a flat spectrum over a definite bandwidth which is of interest for a specific application (for example audio frequency band or radio frequency band).

The physical bandwidth of white noise is limited in practice by various factors such as the mechanism of noise generation, the transmission medium and finite observation capabilities. The finite spectral band implies some correlations between the random variables of the noise process, which significantly increases the mathematical complexity of the problem. A consistent example of finite-band white noise is the Ornstein-Uhlenbeck process, which is addressed in the next section.

2.1.5 Ornstein-Uhlenbeck Noise

The Ornstein-Uhlenbeck (OU) processes belong to a class of finite-band WGN, whose spectral densities are constant in the small frequency region and decrease to zero inversely proportional to the square frequency in the high frequency region. More specifically, OU spectral density has a Lorentzian shape, $S(f) = \sigma^2/(b^2 + 4\pi^2 f^2)$ where σ and b are constants characteristic to the process and f is the frequency (see Fig. 2.5). It is used in modeling various thermal relaxation processes as well as the evolution of exchange rates, bank interest, or prices.

The mathematical description of the OU process can be simply obtained by adding a linear drift term to the FPE characterizing the Wiener process. Thus, the FPE for the transition probability function of the OU noise reads:

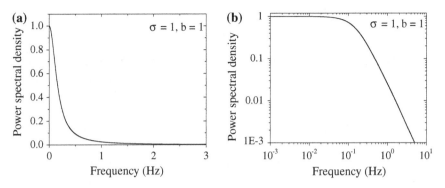

Fig. 2.5 Power spectral density of the Ornstein-Uhlenbeck process on, (**a**) linear scale and, (**b**) Logarithmic scale

$$\frac{\partial}{\partial t}p(x,t|x_0,t_0) = \frac{\partial}{\partial x}[b(x-x_s)p(x,t|x_0,t_0)] + \frac{\sigma^2}{2}\frac{\partial^2}{\partial x^2}p(x,t|x_0,t_0), \qquad (2.24)$$

where b, x_s, and σ are constants known as drift coefficient, stationary average and diffusion coefficient, respectively. The solution is subject to the initial condition $p(x,t|x_0,t_0) = \delta(x-x_0)$ and has to decay to zero as x goes to infinity. In physical terms, OU process can be interpreted as a Brownian particle diffusing in a parabolic potential $U(x)$ with derivative $U'(x) = b(x-x_s)$.

To find the solution of FPE (2.24) let us consider, by simple translation of variables, $t_0 = 0$ and $x_s = 0$. By applying Fourier transformation with respect to x variable, $\tilde{p}(s,t|x_0,0) = \int_{-\infty}^{\infty} p(x,t|x_0,0)e^{isx}dx$, Eq. (2.24) becomes:

$$\frac{\partial}{\partial t}\tilde{p}(s,t|x_0,0) + bs\frac{\partial}{\partial s}\tilde{p}(s,t|x_0,0) = -\frac{\sigma^2}{2}s^2\tilde{p}(s,t|x_0,0), \qquad (2.25)$$

subject to initial condition $\tilde{p}(s,0|x_0,0) = e^{isx_0}$, which can be solved by the method of characteristics. Thus, let us find the characteristic curve from the associated Lagrange-Charpit equations:

$$\frac{dt}{1} = \frac{ds}{bs} = -\frac{2d\tilde{p}}{\sigma^2 s^2 \tilde{p}} \qquad (2.26)$$

By integrating the first equation in (2.26) using the separation of variables and imposing initial condition $s(0) = s_0$, the following solution is found:

$$s(t) = s_0 e^{bt} \qquad (2.27)$$

By plugging this expression for s in the last term of the formula (2.26) and solving the corresponding differential equation with respect to t subject to initial condition $e^{is_0x_0}$, one can use separation of variables to obtain the following solution:

$$\tilde{p}(s_0, t | x_0, 0) = \exp\left(i s_0 x_0 + \frac{\sigma^2 s_0^2}{4b} \left(1 - e^{2bt}\right) \right) \tag{2.28}$$

For the clarity of the previous formula, two notations were used for the exponential function. By substituting s_0 in (2.28) as a function of s and t obtained from (2.27), one arrives at the following solution of the partial differential Eq. (2.25):

$$\tilde{p}(s, t | x_0, 0) = \exp\left(i x_0 s e^{-bt} - \frac{\sigma^2 s^2}{4b} \left(1 - e^{-2bt}\right) \right) \tag{2.29}$$

By performing Fourier inversion, a Gaussian distribution with mean $x_0 e^{-bt}$ and variance $(\sigma^2/2b) (1 - e^{-2bt})$ is obtained. Taking into account the translation of variable used at the beginning of this derivation, the solution of FPE (2.24) is obtained. Thus, the transition probability function of the Ornstein-Uhlenbeck noise are characterized by drift coefficient b, and diffusion coefficient. The average of the stationary noise x_s has the following expression:

$$p(x, t | x_0, t_0) = \frac{1}{\sigma\sqrt{(\pi/b)(1 - e^{-2b(t-t_0)})}} \exp\left[-\frac{b(x - x_s - (x_0 - x_s)e^{-b(t-t_0)})^2}{2\sigma^2(1 - e^{-2b(t-t_0)})} \right],$$

$$\tag{2.30}$$

When t goes to infinity the transition probability exponentially approaches the stationary distribution, which is Gaussian with mean x_s and variance $\sigma^2/2b$. Thus the initial δ-distribution is spread in time (see Fig. 2.6), as happened in the Wiener process, but the standard deviation converges to a finite value when $t \to \infty$. In addition, the distribution center drifts away from the initial condition x_0 to the stationary average x_s.

Let us now compute the autocorrelation function of the OU process by using definition (2.5) and the transition probability function previously derived. Based on the Markovian property of the OU process and assuming that $t_2 > t_1 > t_0$, the autocorrelation function can be rewritten as follows:

Fig. 2.6 Evolution of the transition probability density for an Ornstein-Uhlenbeck process starting at x_0

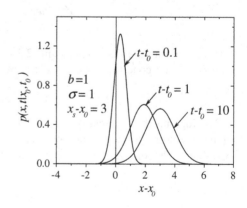

$$\langle X(t_2) \cdot X(t_1)|x_0, t_0\rangle = \iint_{R^2} x_2 x_1 p(x_2, t_2|x_1, t_1) p(x_1, t_1|x_0, t_0) dx_1 dx_2$$

$$= \int_R \langle X(t_2)|x_1, t_1\rangle x_1 p(x_1, t_1|x_0, t_0) dx_1 \qquad (2.31)$$

As it was previously derived, the expression for average at t_2 of an OU process initiated at (x_1, t_1) is $(x_1 - x_s)e^{-b(t_2-t_1)}$, and consequently formula (2.31) becomes:

$$\langle X(t_2) \cdot X(t_1)|x_0, t_0\rangle = e^{-b(t_2-t_1)} \int_R (x_1 - x_s) x_1 p(x_1, t_1|x_0, t_0) dx_1$$

$$= e^{-b(t_2-t_1)} \left[\langle X^2(t_1)|x_0, t_0\rangle - x_s \langle X(t_1)|x_0, t_0\rangle \right] \qquad (2.32)$$

Since the second moment is the sum of the square average and variance, the autocorrelation function of the OU process becomes:

$$\langle X(t_2) \cdot X(t_1)|x_0, t_0\rangle = e^{-b(t_2-t_1)} \left\{ \frac{\sigma^2}{2b} - x_s(x_0 - x_s)e^{-b(t_1-t_0)} + \left[(x_0 - x_s)^2 - \frac{\sigma^2}{2b} \right] e^{-2b(t_1-t_0)} \right\}$$

$$(2.33)$$

Let us observe that the autocorrelation expression is significantly simplified when the initial condition is the stationary average or is considered in the remote past. Actually, the latter is of much more interest from a practical point of view and is coined as the *stationary correlation function*, denoted by $<X(t_2)X(t_1)>_s$. By letting $t_0 \to -\infty$ in formula (2.33), one gets:

$$\langle X(t_2) \cdot X(t_1)\rangle_s = \frac{\sigma^2}{2b} e^{-b|t_2-t_1|} \qquad (2.34)$$

where the absolute value was used in order to account for both $t_2 > t_1$, as considered in the previous derivation, and $t_1 > t_2$. The fact that the autocorrelation function depends only on time difference is characteristic to stationary process. It is natural to require for a stochastic process modeling the noise to be a stationary memoryless (i.e. Markovian) process. If the Gaussian requirement for the distribution function is added then the OU process is the only one that satisfies all these three natural characteristics, as it is proven by the Doob theorem [20].

The power spectral density of the OU process can now be easily obtained as the Fourier transform of the autocorrelation function (2.34) according to the Wiener-Khinchine theorem [16]. Because we deal with an even correlation function, it is enough to compute the Fourier integral on the positive axis. Thus,

$$S(\omega) = 2\text{Re}\left\{ \int_0^\infty \frac{\sigma^2}{2b} e^{-b\tau} e^{-j\omega\tau} d\tau \right\} = \frac{\sigma^2}{b} \text{Re}\left\{ \frac{1}{b + i\omega} \right\} = \frac{\sigma^2}{b^2 + \omega^2} \qquad (2.35)$$

which proves the Lorentzian shape of the OU spectrum mentioned at the beginning of this section and illustrated in Fig. 2.5.

Based on FPE (2.24) for OU processes, the associated Itô stochastic differential equation can be simply written down as:

$$dX(t) = -b[X(t) - x_s]dt + \sigma \cdot dW(t) \qquad (2.36)$$

where $W(t)$ is the Wiener process, b and σ are the drift and, respectively, diffusion coefficients of $X(t)$, while x_s is the average of the stationary process. The process is also subject to the initial condition $X(0) = x_0$. While analytical calculations involving the OU process performed in the book are mainly based on the FPE approach, numerical simulations are using the Itô SDE description as it is next discussed.

By using the finite difference technique and the fact that $W(s + t) = W(s) + N(0, 1)t^{1/2}$, where $N(0, 1)$ is a random variable normally distributed with zero average and unit variance, one obtains the following *approximate* updating formula:

$$x(t + \Delta t) \approx x(t) - b[x(t) - x_s]\Delta t + \sigma \cdot N(0, 1)(\Delta t)^{1/2} \qquad (2.37)$$

Although Eq. (2.37) has often been used in the literature to generate OU processes, it is reliable only when Δt is relatively small. An *exact* updating formula has been derived in [21] by integrating (2.36) and by using the properties of normal variables:

$$x(t + \Delta t) = x(t)e^{-b\Delta t} + \left[\left(\sigma^2/2b\right)(1 - e^{-2b\Delta t})\right]^{1/2}N(0, 1) \qquad (2.38)$$

in which it is assumed that $x_0 = x_s = 0$.

As expected, this updating formula is reduced to (2.37) when $\Delta t << 1/b$. It is noteworthy that Eq. (2.38) splits explicitly the random process into two terms: the first one is the mean and the second one is proportional to the standard deviation of $x(t)$. Since the time step Δt is usually constant, the factors in (2.38) can be computed in advance and stored in order to increase the computational efficiency. This latter approach has been used in our book to generate OU processes numerically. Sample paths of OU are shown in Fig. 2.7.

2.1.6 Brownian Motion in a Double Well-Potential

In this section, the discussion is extended from the Brownian motion in one-well potential reflected by Ornstein-Uhlenbeck process to the Brownian motion in double-well potential which obeys the following Fokker-Planck equation:

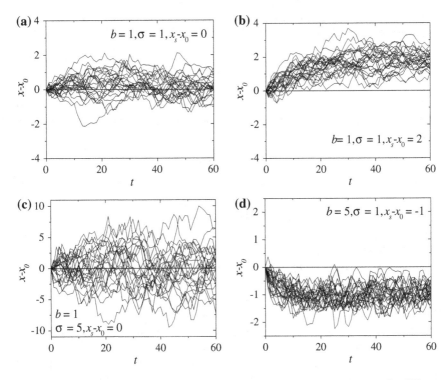

Fig. 2.7 Simulated sample paths of the Ornstein-Uhlenbeck process starting at x_0 for different values of b, σ, x_s

$$\frac{\partial}{\partial t}p(x,t|x_0,t_0) = \frac{\partial}{\partial x}\left[\frac{dU}{dx}(x)p(x,t|x_0,t_0)\right] + \frac{\sigma^2}{2}\frac{\partial^2}{\partial x^2}p(x,t|x_0,t_0), \qquad (2.39)$$

where $U(x)$ denotes a function at least twice differentiable having two minima inside the interval of interest for the problem, as illustrated in Fig. 2.8.

Fig. 2.8 Plot of double-well potential $U(x)$ and the stationary distribution $p_s(x)$

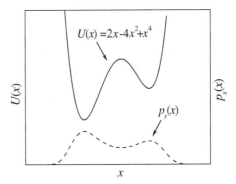

Let us first find the stationary distribution, which is obtained by solving the following differential equation:

$$0 = \frac{d}{dx}\left[\frac{dU}{dx}(x)p_s(x)\right] + \frac{\sigma^2}{2}\frac{d^2p_s(x)}{dx^2} \tag{2.40}$$

Since both the probability function and its derivatives have to approach zero when x goes to infinity, the constant corresponding to the first integration is zero and Eq. (2.40) is equivalent to the following:

$$\frac{dp_s(x)}{dx} = -\frac{2}{\sigma^2}\frac{dU}{dx}(x)p_s(x) \tag{2.41}$$

which can be easily solved using separation of variables.

It is apparent from formula (2.41) that the minima for potential $U(x)$ are maxima for the stationary probability $p_s(x)$ representing metastable states (see also Fig. 2.8). A natural problem to be discussed is the transition between the two metastable states induced by noise. It is intuitively clear that the time needed to pass from one metastable state to another is mostly spent by surmounting the potential barrier between the states. The latter can be seen as the time needed for the Brownian particle initially located in one minimum to escape from the corresponding half-bounded interval ending the maximum point. In order to compute this exit time let us impose an absorbing boundary condition on Eq. (2.39) at the maximum point M, i.e. $p(M,t|x_0, 0) = 0$. The corresponding solution will provide the probability that, at time t, the particle starting at x_0 is still in the first potential well, which will be denoted by $G(x_0, t)$ and has the following expression:

$$G(x_0, t) = \int_{-\infty}^{M} p(x, t|x_0, 0)dx \tag{2.42}$$

In other words, $G(x_0, t)$ represents the tail distribution of the first exit time from the potential well and consequently, the mean first exit time, denoted by $T(x_0)$ can be expressed as follows:

$$T(x_0) = \langle t \rangle = \int_0^{\infty} t\frac{\partial}{\partial t}(1 - G(x_0, t))dt = \int_0^{\infty} G(x_0, t)dt \tag{2.43}$$

where the last equality is obtained using integration by parts.

By taking into account that the transition probability satisfies the backward Fokker-Planck equation as function of initial condition x_0 and time t, $G(x_0, t)$ obeys the following equation:

$$\frac{\partial}{\partial t}G(x_0, t) = \frac{dU}{dx_0}(x_0)\frac{\partial}{\partial x_0}G(x_0, t) + \frac{\sigma^2}{2}\frac{\partial^2}{\partial x_0^2}G(x_0, t) \tag{2.44}$$

subject to the initial condition $G(x_0, 0) = 1$ for all x_0 smaller than M and boundary condition $G(M, t) = 0$ and decays to zero as x_0 goes to minus infinity for all $t > 0$. By integrating this equation over time from 0 to infinity, the equation for the mean first exit time is obtained:

$$\frac{dU}{dx_0}(x_0)\frac{d}{dx_0}T(x_0) + \frac{\sigma^2}{2}\frac{d^2}{dx_0^2}T(x_0) = -1. \tag{2.45}$$

with the boundary conditions $T(M) = 0$ and decays to zero when x_0 goes to minus infinity. It is apparent that Eq. (2.45) is a linear first order differential equation in terms of the derivative of T, so the analytical solution is readily available:

$$\frac{dT}{dx_0}(x_0) = e^{\frac{2U(x_0)}{\sigma^2}}\left(-\frac{2}{\sigma^2}\int_{-\infty}^{x_0} e^{-\frac{2U(x)}{\sigma^2}}dx + c\right) \tag{2.46}$$

where c is an integration constant that is to be determined from the boundary conditions on T. Let us mention that if instead of $-\infty$ is considered a finite left bound with reflective boundary condition, the derivative of T is equal to zero at that point, so the constant c is also zero. By integrating (2.46) and taking into account the boundary conditions $T(M) = 0$, the following closed form expression is obtained for the mean first exit time:

$$T(x_0) = \frac{2}{\sigma^2}\int_{x_0}^{M} e^{\frac{2U(x)}{\sigma^2}}\left(\int_{-\infty}^{x} e^{-\frac{2U(y)}{\sigma^2}}dy\right)dx \tag{2.47}$$

Once $U(x)$ is explicitly given, the expression (2.47) can be further simplified by computing the two integrals. Here, let us consider that diffusion strength σ^2 is relatively small compared to the height of the potential barrier. On the one hand, $\exp[2U(x)/\sigma^2]$ is sharply peaked at $x = M$ so the main contribution to the first integral comes from a close neighborhood of M, where $U(x)$ can be approximated by $U(M)-\beta(x-M)^2$ with β a constant from Taylor approximation formula. On the other hand, $\exp[-2U(x)/\sigma^2]$ is very small near $x = M$ so the inner integral is very slowly varying in the close neighborhood of M significant for the first integral. As a result, the inner integral can be approximated by setting the integral limit $x = M$ and the resulting constant can be removed from inside the first integral. Moreover, the main contribution to this integral comes from the neighborhood of minimum m_1, where $U(x)$ can be approximated by $U(m_1) + \alpha(x-m_1)^2$ with α a constant that comes from the Taylor approximation formula. By taking into consideration all these observations, the mean first exit time of a particle located at metastable state m_1 can be approximated by the following formula:

$$T(m_1) \approx \frac{2}{\sigma^2} \int\limits_{-\infty}^{M} e^{-\frac{2[U(m_1)+\alpha(y-m_1)^2]}{\sigma^2}} dy \int\limits_{m_1}^{M} e^{-\frac{2[U(M)-\beta(x-M)^2]}{\sigma^2}} dx$$

$$\approx \frac{2}{\sigma^2} e^{\frac{2(U(M)-U(m_1))}{\sigma^2}} \int\limits_{-\infty}^{\infty} e^{-\frac{2\alpha(y-m_1)^2}{\sigma^2}} dy \int\limits_{-\infty}^{M} e^{-\frac{2\beta(x-M)^2}{\sigma^2}} dx \tag{2.48}$$

It is relatively easy to show that the first integral gives $\sigma\sqrt{\pi/2\alpha}$ and the second integral gives $\sigma\sqrt{\pi/8\beta}$. In conclusion, when noise strength is relatively small compared to the potential barrier, the escape time can be approximated by the following expression:

$$T(a) \approx \frac{\pi}{2\sqrt{\alpha\beta}} e^{\frac{2(U(b)-U(a))}{\sigma^2}} \tag{2.49}$$

This result is known as Arrhenius formula and has been frequently used in modeling thermal relaxation phenomena, where the noise strength is proportional to the absolute temperature of the system ($\sigma^2/2 = kT$, k is the Boltzmann's constant).

The analytical solutions for the transition probability function of the Brownian motion in double-well potential are much more difficult to find than in the case of one-well potential where Fourier method was effective. These solutions can be obtained in terms of eigenfunctions for FPE (2.39) with the eigenvalues determining the rates of decay to the stationary state [16, 22]. Here, we focus on numerical simulations of the process which are addressed by solving the associated SDE:

$$dX(t) = -\frac{dU}{dx}(X(t))dt + \sigma \cdot dW(t) \tag{2.50}$$

By using the finite difference technique and $W(s + t) = W(s) + N(0, 1)t^{1/2}$, where $N(0,1)$ is a random variable normally distributed with zero average and unit variance, one obtains the following *approximate* updating formula:

$$x(t + \Delta t) \approx x(t) - \frac{dU}{dx}(x(t))\Delta t + \sigma \cdot N(0, 1)(\Delta t)^{1/2} \tag{2.51}$$

Simulated sample paths of Brownian motion in a Landau potential and variants thereof are plotted in Fig. 2.9. The Landau potential is a standard case of double-well potential with $U(x) = -(b/2)x^2 + dx^4$, where constants b and d are positive constants.

2.1.7 Pink (1/f) Noise

Pink noise is a stochastic process with the power spectral density inverse proportional to frequency, also known as "1/f noise". The name of "pink noise" is often extended to any noise with a power spectral density of the form $1/f^\alpha$ where α

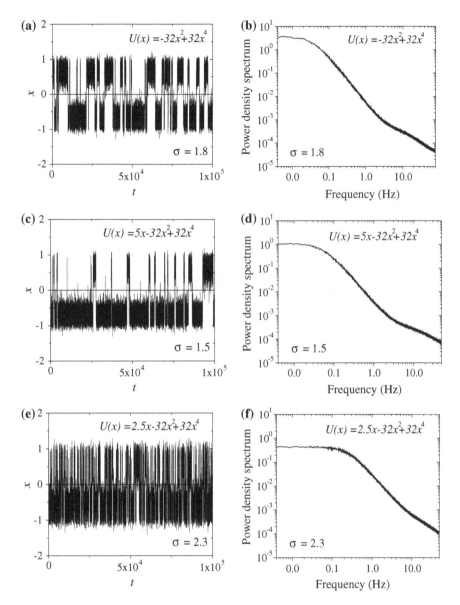

Fig. 2.9 Simulated sample paths of Brownian motion in various potentials $U(x) = (bx_s)x - (b/2)x^2 + dx^4$ and the associated spectra for different values of b, x_s and d

is usually close to 1. Pink noise is considered an intermediate class of noise between the white noise, obtained for $\alpha = 0$, and the Wiener noise featuring a spectrum with $\alpha = 2$.

The pink noise was first observed experimentally by Johnson in 1925 [23] when was trying to measure the noise spectrum in triode vacuum tubes. In addition to the white noise spectrum predicted by Schottky [24] he also observed an unexpected $1/f$ noise at low frequency. In the following years this strange noise appeared again and again in many different electrical devices, as well as in systems from other areas of science and technology, such as biology, astronomy, geophysics, psychology and economics [25, 26]. Several examples are provided in Fig. 2.10.

Although these phenomena are widely spread in nature and their analysis led to more than 1500 scientific publications [27], a unified explanation is still missing. An early approach proposed by Johnson [23] and Shottky [24] was to consider the superposition of various OU relaxation process with different relaxation rates. This model was successful in explaining the pink noise in vacuum tubes but less suited in other cases from the area of electronics. Another idea was to look for diffusion processes as possible origins of pink noise. That was not very difficult from the mathematical point of view but failed at giving consistent physical meaning to the mathematical assumptions used to derive $1/f$ spectrum [28]. Following the Mandelbrot's work [29] on fractals, pink noise has often been associated to fractal phenomena due to its scale invariance, i.e. it does not change if scales of frequency or time are multiplied by a common factor. Moreover, since various nonlinear

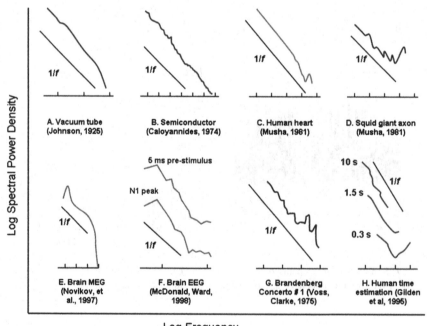

Fig. 2.10 Examples of $1/f$ noises. Curves are illustrative based on data from the indicated sources. Adjacent pairs of *tick marks* on the horizontal axis beneath each figure indicate one decade of frequency. Reprinted with permission from [26]

systems have fractal attractors, many researchers have looked for dynamical systems with complex behavior mimicking a noise process. A pioneering work in this direction has been performed by Bak, Tang and Wiesenfeld who introduced the so-called self organized criticality as an explanation for $1/f$ noise [30, 31]. In conclusion, numerous models have been designed to explain the origin of the pink noise and generate its characteristics. Although no universal approach has been developed, *ad-hoc* models were pretty successful in studying these ubiquitous phenomena.

The simulations of pink noise used in this book are based on the generation of white noise processes and Fourier transforms. If Wiener noise featuring an $1/f^2$ spectrum can be interpreted as the integral of white noise then pink noise featuring an $1/f$ spectrum could be seen as some kind of half-integral of white noise. Let us consider that $\xi(t)$ is a sample path of white noise and compute its Fourier transform. Then, dividing the result by $\omega^{1/2}$ and taking an inverse Fourier transform, one obtains a function of time, denoted by $p(t)$, which defines a sample path of pink noise. This procedure can be mathematically expressed as follows:

$$
\begin{aligned}
p(t) &= \frac{1}{2\pi} \int_{-\infty}^{\infty} \left(\int_{-\infty}^{\infty} \xi(\tilde{t}) e^{j\omega\tilde{t}} d\tilde{t} \right) \omega^{-1/2} e^{j\omega t} d\omega \\
&= \int_{-\infty}^{\infty} \xi(\tilde{t}) \left(\frac{1}{2\pi} \int_{-\infty}^{\infty} \omega^{-1/2} e^{j\omega(\tilde{t}-t)} d\omega \right) d\tilde{t}
\end{aligned}
\tag{2.52}
$$

This equality shows explicitly that pink noise can be constructed as a linear convolution of white noise with a specific kernel (or Green's function) and explains the time correlations in the pink noise. Since there are various types of white noise depending on its probability distribution, a given $1/f$ spectrum can also be associated to a variety of pink noise processes including Gaussian and Laplace noises (Fig. 2.11).

2.1.8 Other Classes of Colored Noise

In general, colored noise is the complementary notion of white noise including noises with flat spectrum only on a finite frequency band and noises with non-flat spectrum. In other words, colored noise spikes are correlated to each other. As it was previously mentioned, white noise bears a physical inconsistency since it requires infinite energy. So practically, all real noises are colored to some degree and pure white noise is only used in theoretical analyses due to its simplicity.

Wiener process featuring $1/f^2$ spectrum, pink noise characterized by $1/f$ spectrum, and Ornstein-Uhlenbeck noise with its Lorentzian spectrum are most common models of colored noise. However, from case to case there is a large variety of colored noises, so modeling noise with arbitrary spectrum is desired.

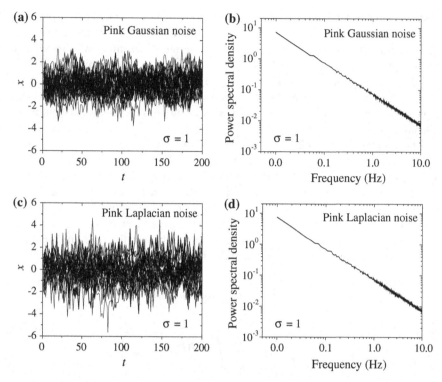

Fig. 2.11 Simulated sample paths of pink noise (*left*) and the associated spectra (*right*) for Gaussian and Laplacian distribution

Our approach generalizes the technique used in the previous section for simulating pink noise. Thus, let us consider an arbitrary positive frequency function $f(\omega)$ sought as the noise spectrum. One first generates numerically an IID process $\xi(t)$ in the time domain as it is done in the white noise case. This process is then converted to the frequency domain by using standard FFT techniques. In order to obtain the desired colored noise $c(t)$ one has to multiply the flat spectrum of the converted signal by the chosen function and convert the signal back to the time domain. This procedure can be mathematically expressed as follows:

$$c(t) = \frac{1}{2\pi} \int_{-\infty}^{\infty} \left(\int_{-\infty}^{\infty} \xi(\tilde{t}) e^{j\omega\tilde{t}} d\tilde{t} \right) \sqrt{f(\omega)} e^{j\omega t} d\omega$$

$$= \int_{-\infty}^{\infty} \xi(\tilde{t}) \left(\frac{1}{2\pi} \int_{-\infty}^{\infty} \sqrt{f(\omega)} e^{j\omega(\tilde{t}-t)} d\omega \right) d\tilde{t}$$

$$(2.53)$$

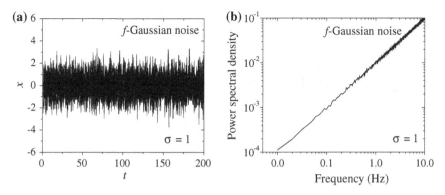

Fig. 2.12 Sample paths of blue noise (*left*) and the associated spectra (*right*)

It should be noted that the computational cost for the generation of the colored noise is relatively small and depends on n as $O(n \log n)$ where n is the length of the signal.

In Fig. 2.12a sample paths of colored noise with a spectrum directly proportional to f (represented in Fig. 2.12b) are generated from white Gaussian noise by using the algorithm previously described. This type of noise is known as *blue noise* and is often detected and used in image processing. Efficient algorithms for dithering were developed by using blue noise. It was found that retina cells are arranged in blue-noise-type pattern which generates a good visual resolution [32]. Simulated sample paths of a colored noise with a spectrum directly proportional to f^2 are plotted in Fig. 2.13a and the corresponding averaged spectrum is represented in Fig. 2.13b. This is known as violet noise and can be seen as a derivative of white noise. It is apparent that infinite-band blue of violet noise also require infinite energy, so such noises can only exist on a finite band. Figure 2.14b shows a finite band spectrum with triangular shape which has violet part and a Brown part. The sample paths associated to this spectrum are represented on the left part of that figure.

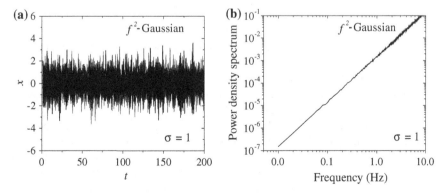

Fig. 2.13 Sample paths of violet noise (*left*) and the associated spectra (*right*)

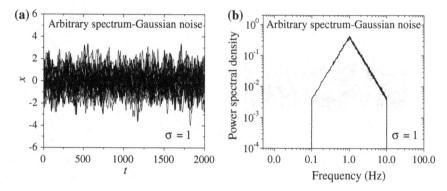

Fig. 2.14 Sample paths of colored noise (*left*) with finite-band triangular spectrum represented on the (*right*) figure

Many other classes of noise can be defined based on the characteristics of their spectra but they are much less encountered in hysteretic systems and consequently, they are not addressed in this book. However, they can be easily generated and applied to specific applications by using the *Noise Module* of *HysterSoft* and by following the procedure presenting above.

In conclusion, most common noise models in hysteretic systems have been presented along with the main numerical techniques used in this book for noise simulations. It has been emphasized that noise may also play a constructive role in nonlinear systems in opposition to the general image of noise as nuisance. Regardless of their positive or negative roles, it is clear that a physical system is influenced by internal of external noise leading to a stochastic behavior of the system output. The next part of the Chapter is devoted to the theory of stochastic processes defined on graphs, which proved to be naturally suited to the stochastic analysis of the hysteretic system outputs.

2.2 Stochastic Processes Defined on Graphs

This section is devoted to introducing the theory of stochastic processes defined on graphs that was recently developed by Freidlin and Wentzell. Their papers [33, 34] are used as a guide for presenting the basic concepts of this theory. First, several definitions and general properties of stochastic processes are discussed, stressing the relation between transition probabilities of Markov processes and semigroups of contractions. This relation allows the characterization of diffusion processes defined on a graph, which is addressed in the second part of this section and is later applied to the analytical study of hysteretic systems with stochastic input. Readers without a background in measure theory and functional analysis might find difficult to understand this theoretical construction so they can pass directly to the Sect. 2.3.3 part where the theory is applied to the case of Orstein-Uhlenbeck process defined on a graph.

2.2.1 General Properties of Diffusion Processes

Consider a probability space $\{\Omega, F, \mathrm{P}\}$ where Ω is the set of outcomes known as *sample space*, F is a collection of subsets of Ω which forms σ-algebra, and P is the probability measure returning the probability of a specific event in Ω. In addition, let us consider two real intervals X (phase space) and T (time interval). Let us remind that a *stochastic process* is a family of random variables $\{X(t)\}$, $t \in T$, defined on Ω with values in X. For each fixed $\omega \in \Omega$ a function $x{:}T \to X$, is obtained as $x(t) = X(t)(\omega)$ and is known as the *trajectory* or *sample path* of the process $X(t)$. A stochastic process is called (*right*) *continuous* if "almost all" of its trajectories are (right) continuous, where "almost all" means a property valid on a subset of Ω which has measure 1.

The collection of probability distribution functions $f_{t_1 t_2 \ldots t_r}$ of random variables $(X(t_1), X(t_2), \ldots X(t_r))$ for any natural number r and for any $t_1, t_2, \ldots t_r \in T$ is known as the *finite-dimensional family of distributions* of process $X(t)$. In general, the finite-dimensional family of distributions is not uniquely defining a stochastic process, but there is a large class of stochastic processes[1] for which it determines "almost" unique a continuous stochastic process. All processes considered in this section satisfy this property.

A homogeneous Markovian process with respect to a non-decreasing system of σ-algebras $N_t \subset F$, where $t \in T = [0, \infty)$, is by definition a couple formed by a stochastic process $X(t)$ and a collection of probability measures p_x, $x \in X$, on $\{\Omega, F\}$, which satisfy the following conditions:

(1) for any t, random variable $X(t)$ is measurable with respect to σ-algebra N_t;
(2) for any t and any Borel set $\Gamma \subset X$, $P(t, \Gamma|x) = P_x(X(t) \in \Gamma)$ is a Borel function with respect to variable x;
(3) $P(0, X \backslash \{x\}|x) = 0$;
(4) if $t, u \in T$, $t \leq u$, $x \in X$, and $\Gamma \subset X$ is a Borel set, then equality

$$P_x\{X(u) \in \Gamma|N_t\} = P(u - t, \Gamma|x)$$

is satisfied almost certainly with respect to the measure P_x, where $P_x\{A|N_t\}$ represents the conditional probability of the event in relation to σ-algebra N_t;
(5) if $u \geq 0$ then for each $\omega \in \Omega$ exists $\omega' \in \Omega$ such that the equality

---

[1] See Kolmogorov continuity theorem [34].

$$(X(t + u))(\omega') = (X(t))(\omega)$$

is satisfied for all t.

Intuitively, Markov processes can be interpreted as stochastic processes without memory. The definition considered that the process $X(t)$ is defined for any $t \in [0, \infty)$. However it should be noted that many problems lead to processes $(X(\cdot))(\omega)$ that are defined only for a finite range $[0, \xi(\omega)]$, where random variable $\xi(\omega)$ is called *terminal time*. Since no such processes are used in this book, we have simplified to a certain extent this definition.

The notion of homogeneity for a Markov process is directly related to property (5), which implies the invariance of the set of Markov process trajectories at the translation of time. Function $P(t, \Gamma | x)$ is called the *transition probability function* of the Markov process and determines, to a certain degree of equivalence,[2] the stochastic process. Thus, the properties and proper analysis of Markov processes are often reduced to the properties and analysis of transition probabilities. For the rigorous foundation of this schematic presentation the reader may consult the monographs by Dynkin [34] and Mandl [35].

A Markov process can be associated to a semigroup of contractions S_t acting on the Banach space **B** of bounded and measurable functions on X endowed with the *supremum* norm. It is defined by the formula:

$$(S_t f)(x) = \int_X f(y) P(t, dy | x) \tag{2.54}$$

The infinitesimal generator A of this semigroup, and hence of the associated Markov process, is defined by the following formula:

$$Af = \lim_{t \to 0} \frac{S_t f - f}{t} \tag{2.55}$$

where convergence is considered the supremum norm. In general, A cannot be defined for all elements of **B**. A special problem related to the definition of infinitesimal generator is the boundary condition $(x \in Fr\{X\})$. Thus, different types of behavior of Markov process at phase space boundary correspond to different boundary conditions for the functions f defining the domain $D(A)$ of the infinitesimal generator. For each function $f \in D_A$, the function $u_t(x) = S_t f(x)$ is the unique (bounded) solution of the following Cauchy problem:

$$\frac{\partial u_t(x)}{\partial t} = Au_t(x), \quad \lim_{t \to 0} u_t(x) = f(x) \tag{2.56}$$

If the transition probability of stochastic process is continuous then the semigroup S_t (and hence the infinitesimal generator A) uniquely determines this transition

---

[2] See Theorem 3.2, page 85, Ref. [34].

probability and all finite-dimensional family of distributions for the Markov process.

An important class of Markov processes is composed of diffusion processes, which requires some additional restrictions on the transition probability functions. Let us consider that, for each $x \in X$ the following limits exist:

$$\lim_{t \to 0} t^{-1} \left[1 - \int_X P(t, dy|x) \right] = 0 \qquad (2.57)$$

$$\lim_{t \to 0} t^{-1} \left[1 - \int_X (y - x)P(t, dy|x) \right] = b(x) \qquad (2.58)$$

$$\lim_{t \to 0} t^{-1} \left[1 - \int_X (y - x)^2 P(t, dy|x) \right] = \sigma^2(x) \qquad (2.59)$$

where the function $b(x)$ is known as the *drift coefficient*, while $\sigma(x) \geq 0$ as *diffusion coefficient* of the transition probability, and hence of the associated Markov process. A Markov process satisfying these conditions is called *diffusion process*. Note that the action on the class $C^2(X)$ functions of the infinitesimal generator associated to a diffusion process is given by:

$$Af = \lim_{t \to 0} t^{-1} \left[\int_X f(y)P(t, dy|x) - f(x) \right] = \frac{1}{2}\sigma^2(x)\frac{d^2 f}{dx^2}(x) + b(x)\frac{df}{dx}(x) \qquad (2.60)$$

which clarifies to some extent, the conditions imposed to define diffusion processes. This relationship also suggests a deep connection between the diffusion processes and elliptical differential operators.

Differential operator G defined by the formula:

$$G = \frac{1}{2}\sigma^2(x)\frac{\partial^2}{\partial x^2} + b(x)\frac{\partial}{\partial x} \qquad (2.61)$$

is known as *differential generator* of the diffusion process. In some conditions of weak regularity imposed on the drift and diffusion coefficients, the diffusion process is uniquely determined by its differential generator, meaning that any two processes with the same differential generator generate the same distribution in the space of trajectories (sample paths).

If there is a positive real constant K such that for any $x, y \in X$ these coefficients satisfy:

- *Lipschitz condition* : $|b(x, t) - b(y, t)| + |\sigma(x, t) - \sigma(y, t)| \leq K|x - y|$;
- *Growth condition* : $|b(x, t)|^2 + |\sigma(x, t)|^2 \leq K\left(1 + |x|^2\right)$

then exits a unique fundamental solution, denoted by $\rho(t, x, y)$, of equation $\partial u/\partial t = Gu$ satisfying the appropriate initial and boundary conditions. This solution is precisely the transition probability density associated to the given diffusion process. Thus,

$$P(t, \Gamma | x) = \int_{\Gamma} \rho(t, y | x) dy \qquad (2.64)$$

Equation $\partial \rho/\partial t = G\rho$ is called backward Kolmogorov equation of the diffusion process. In the end, let us note that a stochastic process is called *conservative* if $P(t, X | x) = 1$ for any t and x.

2.2.2 Diffusion Processes Defined on Graphs

The theory of stochastic processes on a graph has been recently developed by Freidlin and Wentzell [36]. This theory was first applied to the study of random perturbations of Hamiltonian dynamical systems [33, 36]. Then, it has been realized that this mathematical technique is naturally suitable for the analysis of noise in hysteretic systems [37–41]. In this section, we give a short description of diffusion processes on a graph based on the previously cited references and adapted to the problems of interest in this book. In the end, the initial-boundary value problem for the transition probability density of the diffusion process $Z(t)$ defined on a graph is derived.

Consider a connected graph Z with vertices $V_1,\ldots, V_m$ and edges $E_1,\ldots, E_n$ (see an example in Fig. 2.15). On each edge E_j is taken a coordinate x_j and the distance between two points on the graph is the length of the shortest path connecting those two points measured using the coordinate x_j. Note that the definition of Markov processes given in the previous section can be generalized easily for the case when phase space is considered to be the graph Z, by replacing the symbol X representing a real interval with the symbol Z representing the convex graph. Similarly a semigroup of contractions S_t and an infinitesimal operator A, is associated to the Markov process.

Fig. 2.15 The graph on which the diffusion process is defined

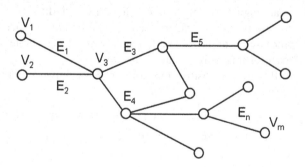

Several edges can meet at a vertex V_k; we will write $E_j \sim V_k$ if the edge E_j has the vertex V_k as its end. For a function $f : Z \to R$ and a segment $E_j \sim V_k$, $(df/dx_j)(V_k)$ denotes the derivative function f with respect to the coordinate x_j considered towards inside of the edge E_j. A diffusion process $X^j(t)$ is associated with each edge E_j and is defined by the differential generator:

$$G_j = b_j(x_j)\frac{\partial}{\partial x_j} + \frac{\sigma_j^2(x_j)}{2}\frac{\partial^2}{\partial x_j^2} \qquad (2.65)$$

where, b_j and σ_j are continuous functions that satisfy Lipshitz condition (2.62) and growth condition (2.63).

For any nonnegative constants α_k and χ_{kj}, with $\alpha_k + \sum_{j:E_j \sim V_k} \chi_{kj} > 0$ for $k = 1,...,m$, one can define an operator A as:

$$Af(\mathbf{z}) = G_j f(\mathbf{z}), \quad \text{pentru } \mathbf{z} \in E_j \qquad (2.66)$$

for all functions f from $C(Z)$ that satisfy the following conditions:

1. f is twice continuously differentiable inside the edges E_j;
2. if $E_j \sim V_k$ then $\lim_{\mathbf{z} \to V_k, \mathbf{z} \in E_j} G_j f(\mathbf{z})$ exists and is independent of j; this limit will be denoted by $Gf(V_k)$;
3. for each vertex V_k

$$\alpha_k Af(V_k) = \sum_{j:E_j \sim V_k} \chi_{kj}\frac{\partial f}{\partial x_j}(V_k); \qquad (2.67)$$

these conditions at the vertices will be further called "gluing" conditions.

The following result has been proven by Freidlin and Wentzell in Ref. [36]:

Theorem *The operator A defined above is the infinitesimal generator of a continuous semigroup of linear operators on $C(Z)$ corresponding to a continuous conservative Markov process $\mathbf{Z}(t)$ on the graph Z.*

Conversely, let $\mathbf{Z}(t)$ be a continuous conservative Markov process defined on the graph Z whose trajectories coincide, up to the exit from the edge E_j, with the diffusion process generated by the operator G_j defined by formula (2.65) and whose associated semigroup of linear operators leads $C(Z)$ into itself. Then there exist unique positive constants χ_{kj} and α_k satisfying $\alpha_k + \sum_{j:E_j \sim V_k} \chi_{kj} > 0$ such that the infinitesimal generator associated to the Markov process $\mathbf{Z}(t)$ is the operator A defined above.

Intuitively, constants α_k describe how much time the process spends in V_k and constants χ_{kj}, are (roughly speaking) proportional to the probabilities that the process will "move" from vertex V_k along the edges E_j.

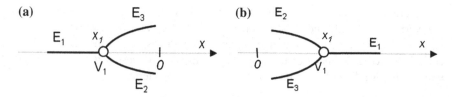

Fig. 2.16 Typical graph configurations used in the analysis of hysteretic systems

For the models used in the next chapters the following facts can be established:

- Since the process has no delay at the vertices, $\alpha_k = 0$ for all k.
- In each interior vertex of the graph there are connected three edges and there is zero probability that the process will move from the vertex to one edge (so the associated χ_{kj} coefficient is also zero) while random motion along the other two are equally probable (so the associated χ_{kj} coefficients are equal to one).

The graphs shown in Fig. 2.16 represent typical vertex connections for the problems discussed in this book. For these graphs there is zero probability to move from V_1 along the edge E_3 and equal probability to move from V_1 along the edges E_1 and E_2. Consequently, $\chi_{13} = 0$, $\chi_{11} = 1$, $\chi_{12} = 1$ and taking into account the coordinates on each edges, the following gluing condition can be derived for vertex V_1:

$$\frac{df_{E_1}}{dx}(x_1) = \frac{df_{E_2}}{dx}(x_1) \tag{2.68}$$

Similar assertions are valid for each interior vertex and analogous interface conditions can be derived.

The next task is to specify the partial differential equations for the transition probability density $\rho(t, \mathbf{z}|\mathbf{z}_0, 0)$ corresponding to the Markov process $\mathbf{Z}(t)$. The following notation for the transition probability density is used on each edge E_j:

$$\rho^{(j)}(t, x|\mathbf{z}_0, 0) = \rho(t, \mathbf{z}|\mathbf{z}_0, 0)|_{\mathbf{z} \in E_j} \tag{2.69}$$

According to the theory of Markovian processes, the following equality is valid for $\rho^{(j)}$:

$$\sum_{j=1}^{k} \int_{E_j} f \frac{\partial \rho^{(j)}}{\partial t} dx = \sum_{j=1}^{k} \int_{E_j} (G_j f) \rho^{(j)} dx \tag{2.70}$$

Integrating by parts in formula (2.70) and taking into account the interface conditions presented above and the fact that f can be chosen arbitrary in the domain of the infinitesimal operator A_Y, one finds that the transition probability density $\rho(t, \mathbf{z}|\mathbf{z}_0, 0)$ satisfies the following forward Kolmogorov equation:

$$\frac{\partial \rho_j(x,t|\mathbf{z}_0,0)}{\partial t} + L_j \rho_j(x,t|\mathbf{z}_0,0) = 0 \text{ on each edge } E_j \tag{2.71}$$

where

$$\hat{L}_j \rho = -\frac{1}{2}\frac{\partial^2}{\partial x^2}\left(\sigma_j^2(x)\rho\right) + \frac{\partial}{\partial x}\left(b_j(x)\rho\right) \tag{2.72}$$

and "vertex" type boundary conditions which express the continuity of the transition probability density at the transition between two edges (for example, edges E_1 and E_2 from the graphs shown in Fig. 2.16) and zero boundary condition imposed on the third edge connected at that vertex. On the other hand, the probability current has to be conserved at each vertex. For vertex V_1 from Fig. 2.16, these conditions can be expressed analytically as follows:

$$\rho_1(x_1,t|\mathbf{z}_0,0) = \rho_2(x_1,t|\mathbf{z}_0,0), \quad \rho_3(x_1,t|\mathbf{z}_0,0) = 0,$$
$$\frac{\partial \rho_2}{\partial x}(x_1,t|\mathbf{z}_0,0) + \frac{\partial \rho_3}{\partial x}(x_1,t|\mathbf{z}_0,0) = \frac{\partial \rho_1}{\partial x}(x_1,t|\mathbf{z}_0,0). \tag{2.73}$$

and the transition probability decays to zero at the external noise of the graphs. In addition, the initial conditions is $\rho(\mathbf{z},0|\mathbf{z}_0,0) = \delta_{\mathbf{z}\mathbf{z}_0}$.

2.2.3 Examples: Ornstein-Uhlenbeck Processes on Graphs

In this section, it is shown how the theory of stochastic processes on graphs can be applied to specific problems. Examples of Ornstein-Uhlenbeck processes defined on graphs are presented and the explicit forms of the initial-boundary value problems for the associated transition probability function are derived and solved.

The corresponding transition probability function $\rho(t,\mathbf{z}|\mathbf{z}_0,0)$ can be expressed by its four components $\rho_i(t,x|x_0,0)$ corresponding to the four edges E_i of the graph represented in Fig. 2.17 and defined on the following intervals:

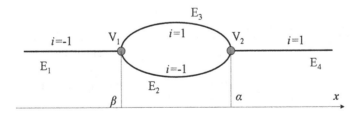

Fig. 2.17 The graph on which the Ornstein-Uhlenbeck process is defined

$\rho_1(t, x|x_0, 0)$ is defined for $x \in (-\infty, \beta)$

$\rho_2(t, x|x_0, 0)$ is defined for $x \in (\beta, \alpha)$

$\rho_3(t, x|x_0, 0)$ is defined for $x \in (\beta, \alpha)$ (2.74)

$\rho_4(t, x|x_0, 0)$ is defined for $x \in (\alpha, \infty)$

Since $\rho_i(t, x|x_0, 0)$ are associated to Ornstein-Uhlenbeck processes (2.89) on these intervals, they are the solutions of the corresponding Fokker–Planck equations defined on the intervals given in formula (2.74):

$$\frac{\partial}{\partial t}\rho_i(x, t|x_0, 0) = \frac{\partial}{\partial x}[b(x - x_s)\rho_i(x, t|x_0, 0)] + \frac{\sigma^2}{2}\frac{\partial^2}{\partial x^2}\rho_i(x, t|x_0, 0), \qquad (2.75)$$

where x_0 is the coordinate of initial point $\mathbf{z}_0$ located on the edge E_{i_0}. The solutions of Eqs. (2.75) are subject to the initial condition $\rho_i(x, t_0|x_0, t_0) = \delta_{ii_0}\delta(x - x_0)$ and to the following "vertex" boundary conditions:

$$\rho_1(\beta^-, t|x_0, 0) = \rho_2(\beta^+, t|x_0, 0)$$
$$\rho_3(\beta^+, t|x_0, 0) = 0$$
$$\rho_3(\alpha^-, t|x_0, 0) = \rho_4(\alpha^+, t|x_0, 0) \qquad (2.76)$$
$$\rho_2(\alpha^-, t|x_0, 0) = 0$$

$$\frac{\partial \rho_1}{\partial x}(\beta^-, t|x_0, 0) = \frac{\partial \rho_2}{\partial x}(\beta^+, t|x_0, 0) + \frac{\partial \rho_3}{\partial x}(\beta^+, t|x_0, 0)$$

$$\frac{\partial \rho_4}{\partial x}(\alpha^+, t|x_0, 0) = \frac{\partial \rho_2}{\partial x}(\alpha^-, t|x_0, 0) + \frac{\partial \rho_3}{\partial x}(\alpha^-, t|x_0, 0)$$

while $p_1(x, t|x_0, 0)$ and $p_4(x, t|x_0, 0)$ have to decay to zero as x goes to minus infinity and plus infinity, respectively.

In order to solve these initial boundary value problems let us observe that the sum of these components $\hat{\rho}(x, t|x_0, 0)$ defined in Eq. (2.77) satisfies Eq. (2.75) for all real values of x except α and β, while vertex boundary conditions prove the continuity and differentiability of this function at α and β and zero decays at $\pm\infty$.

$$\hat{\rho}(x, t|x_0, 0) = \begin{cases} \rho_1(x, t|x_0, 0) & \text{for } x \in (-\infty, \beta) \\ \rho_2(x, t|x_0, 0) + \rho_3(x, t|x_0, 0) & \text{for } x \in (\beta, \alpha) \\ \rho_4(x, t|x_0, 0) & \text{for } x \in (\alpha, \infty) \end{cases} \qquad (2.77)$$

By continuity extension, it is clear that $\hat{\rho}(x, t|x_0, 0)$ satisfies Eq. (2.75) on the entire real axes subject to initial condition $\delta(x - x_0)$ and zero decays at infinity as boundary conditions. Consequently, $\hat{\rho}(x, t|x_0, 0)$ is the standard time-dependent transition probability function of the OU process, which was found in Sect. 1.1.5 to have expression (2.30). As a result,

$$\frac{\sqrt{b}}{\sqrt{\pi\sigma^2(1-e^{-2bt})}}e^{-\frac{b(x-x_s-(x_0-x_s)e^{-bt})^2}{2\sigma^2(1-e^{-2bt})}} = \begin{cases} \rho_1(x,t|x_0,0), & x \in (-\infty,\beta) \\ \rho_2(x,t|x_0,0) + \rho_3(x,t|x_0,0), & x \in (\beta,\alpha) \\ \rho_4(x,t|x_0,0), & x \in (\alpha,\infty) \end{cases}$$

$$(2.78)$$

The transition probability functions of the OU process is completely defined by this formula on the edges E_1 and E_4 of the graph. In addition, the sum of the two transition probability functions corresponding to edges E_2 and E_3 is determined, so only one of them is left to be found in order to solve completely the problem. Let us consider that $i_0 \neq 2$ and choose ρ_2 to be found, otherwise choose ρ_3. Function ρ_2 is the solution of Eq. (2.75) on the interval (β,α) subject to the initial condition $\rho_2(x,0|x_0,0) = 0$ and to the boundary conditions:

$$\rho_2(\beta,t|x_0,0) = \frac{\sqrt{b}}{\sigma\sqrt{\pi(1-e^{-2bt})}}e^{-\frac{b(\beta-x_0)^2}{2\sigma^2(1-e^{-2bt})}}, \quad \rho_2(\alpha,t|x_0,0) = 0 \qquad (2.79)$$

By using Laplace transformation $\tilde{\rho}_2(x,s|x_0,0) = \int_0^\infty e^{-st}\rho_2(x,t|x_0,0)\,dt$ Eq. (2.75) becomes:

$$\frac{\sigma^2}{2}\frac{\partial^2\tilde{\rho}_2}{\partial x^2}(x,s|x_0,0) + b(x-x_s)\frac{\partial\tilde{\rho}_2}{\partial x}(x,s|x_0,0) + (b-s)\tilde{\rho}_2(x,s|x_0,0) = 0 \quad (2.80)$$

which can be solved in terms of special mathematical functions. Hence by considering $\hat{\rho}_2(x,s|x_0,0) = \tilde{\rho}_2(x,s|x_0,0)\exp(b(x-x_s)^2/\sigma^2)$, one obtains the following equation:

$$\frac{\sigma^2}{2}\frac{\partial^2\hat{\rho}_2}{\partial x^2}(x,s|x_0,0) - \left[\frac{b(x-x_s)^2}{2\sigma^2} + s - \frac{b}{2}\right]\hat{\rho}_2(x,s|x_0,0) = 0 \qquad (2.81)$$

This equation has two linearly independent solutions, known as parabolic cylinder functions $U\left(\frac{s}{b}-\frac{1}{2},\frac{\sqrt{2b}}{\sigma}(x-x_s)\right)$ and $V\left(\frac{s}{b}-\frac{1}{2},\frac{\sqrt{2b}}{\sigma}(x-x_s)\right)$ [42]. Consequently, the solution can be expressed as linear combination of U and V with the coefficients, dependent of s, determined from the boundary conditions. In conclusion, a closed form analytical expression for the transition probability ρ_2 can be found in terms of inverse Laplace transforms of the parabolic cylinder functions.

Much simpler analytical results can be found for the stationary distributions. By taking $t \to \infty$ in expression (2.78), one obtains:

$$\hat{\rho}^{st}(x) = \frac{\sqrt{b}}{\sigma\sqrt{\pi}}e^{-\frac{b(x-x_s)^2}{2\sigma^2}} = \begin{cases} \rho_1^{st}(x), & x \in (-\infty,\beta) \\ \rho_2^{st}(x) + \rho_3^{st}(x), & x \in (\beta,\alpha) \\ \rho_4^{st}(x), & x \in (\alpha,\infty) \end{cases} \qquad (2.82)$$

while ρ_2^{st} has to satisfy the equation:

$$\frac{\sigma^2}{2}\frac{\partial^2}{\partial x^2}\rho_2^{st}(x) + \frac{\partial}{\partial x}\left[b(x - x_s)\rho_2^{st}(x)\right] = 0 \tag{2.83}$$

and boundary conditions:

$$\rho_2^{st}(\beta) = \sqrt{\frac{b}{\pi\sigma^2}}e^{-\frac{b(\beta-x_s)^2}{2\sigma^2}}, \quad \rho_2^{st}(\alpha) = 0 \tag{2.84}$$

It is known that the general solution of linear differential Eq. (2.83) has the following form:

$$\rho_2^{st}(x) = e^{-\frac{b(x-x_s)^2}{2\sigma^2}}\left(c\int_x^\alpha e^{\frac{b(y-x_s)^2}{2\sigma^2}}dy + d\right) \tag{2.85}$$

where c and d are constants that can be found from boundary conditions (2.84). The null-condition at $x = \alpha$ implies $d = 0$, while the condition at $x = \beta$ leads to:

$$c = \sqrt{\frac{b}{\pi\sigma^2}}\left(\int_\beta^\alpha e^{\frac{b(y-x_s)^2}{2\sigma^2}}dy\right)^{-1} \tag{2.86}$$

In conclusion, the stationary probability function of the Ornstein-Uhlenbeck process defined on graph Z has the following expression, while a sample obtained for a noise input characterized by $b = 1$, $\sigma = 1$, and $x_s = -0.5$, and vertex coordinates $\beta = -1$ and $\alpha = 1$ is plotted in Fig. 2.18:

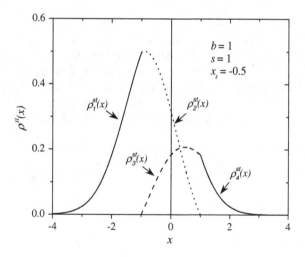

Fig. 2.18 Stationary probability components for an Ornstein-Uhlenbeck process defined on graph Z

$$\rho_1^{st}(x) = \frac{\sqrt{b}}{\sigma\sqrt{\pi}} e^{-\frac{b(x-x_s)^2}{2\sigma^2}}, \quad x \in (-\infty, \beta)$$

$$\rho_2^{st}(x) = \frac{\sqrt{b}}{\sigma\sqrt{\pi}} e^{-\frac{b(x-x_s)^2}{2\sigma^2}} \left(\int_x^\alpha e^{\frac{b(y-x_s)^2}{2\sigma^2}} dy \right) \left(\int_\beta^\alpha e^{\frac{b(y-x_s)^2}{2\sigma^2}} dy \right)^{-1}, \quad x \in (\beta, \alpha)$$

$$\rho_3^{st}(x) = \frac{\sqrt{b}}{\sigma\sqrt{\pi}} e^{-\frac{b(x-x_s)^2}{2\sigma^2}} \left(\int_\beta^x e^{\frac{b(y-x_s)^2}{2\sigma^2}} dy \right) \left(\int_\beta^\alpha e^{\frac{b(y-x_s)^2}{2\sigma^2}} dy \right)^{-1}, \quad x \in (\beta, \alpha)$$

$$\rho_4^{st}(x) = \frac{\sqrt{b}}{\sigma\sqrt{\pi}} e^{-\frac{b(x-x_s)^2}{2\sigma^2}}, \quad x \in (\alpha, \infty)$$

(2.87)

An approximation of the transition probability function defined by (2.75) and (2.76) can be obtained by replacing the stationary distribution $\hat{\rho}^{st}(x)$ of the Orstein-Uhlenbeck process on the real line with the transition probability function $\hat{\rho}(x, t|x_0, 0)$ of the Orstein-Uhlenbeck process on the real line:

$$\rho_1(x, t|x_0, 0) = \frac{\sqrt{b}}{\sqrt{\pi\sigma^2(1 - e^{-2bt})}} e^{-\frac{b(x-x_s-(x_0-x_s)e^{-bt})^2}{2\sigma^2(1-e^{-2bt})}}, \quad x \in (-\infty, \beta)$$

$$\rho_2(x, t|x_0, 0) \approx \frac{\sqrt{b}}{\sqrt{\pi\sigma^2(1 - e^{-2bt})}} e^{-\frac{b(x-x_s-(x_0-x_s)e^{-bt})^2}{2\sigma^2(1-e^{-2bt})}} \int_x^\alpha e^{\frac{b(y-x_s)^2}{2\sigma^2}} dy \left(\int_\beta^\alpha e^{\frac{b(y-x_s)^2}{2\sigma^2}} dy \right)^{-1}, \quad x \in (\beta, \alpha)$$

$$\rho_3(x, t|x_0, 0) \approx \frac{\sqrt{b}}{\sqrt{\pi\sigma^2(1 - e^{-2bt})}} e^{-\frac{b(x-x_s-(x_0-x_s)e^{-bt})^2}{2\sigma^2(1-e^{-2bt})}} \int_\beta^x e^{\frac{b(y-x_s)^2}{2\sigma^2}} dy \left(\int_\beta^\alpha e^{\frac{b(y-x_s)^2}{2\sigma^2}} dy \right)^{-1}, \quad x \in (\beta, \alpha)$$

$$\rho_4(x, t|x_0, 0) = \frac{\sqrt{b}}{\sqrt{\pi\sigma^2(1 - e^{-2bt})}} e^{-\frac{b(x-x_s-(x_0-x_s)e^{-bt})^2}{2\sigma^2(1-e^{-2bt})}}, \quad x \in (\alpha, \infty)$$

(2.88)

Samples of these transitions probability functions obtained for a noise input characterized by $b = 1$, $\sigma = 1$, $x_s = -0.5$, $x_0 = 0$ and vertex coordinates $\beta = -1$ and $\alpha = 1$ are plotted in Fig. 2.19 at selected instants of time.

In the next chapters, it is proven that the stochastic analysis of various hysteretic systems driven by OU processes can be reduced to the analysis of OU processes defined on graphs and the solutions derived here will be useful in expressing the stochastic characteristics of the output.

In the second example, we consider the same graph Z represented in Fig. 2.17 but the Ornstein-Uhlenbeck processes $X^i(t)$ on each edge are governed by different differential generators:

$$G_i = -b\left(x - x_s^i\right)\frac{\partial}{\partial x} + \frac{\sigma^2(x)}{2}\frac{\partial^2}{\partial x^2}$$

(2.89)

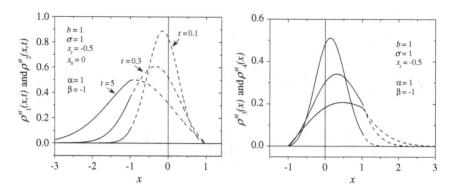

Fig. 2.19 Evolution of the transition probability components for an Ornstein-Uhlenbeck process defined on graph Z

where $x_s^1 = x_s^2 = \tilde{x}_s$ and $x_s^3 = x_s^4 = x_s$. While on edges E_1 and E_2, the process can be interpreted as a Brownian motion in a parabolic potential defined on $(-\infty, \alpha]$ reaching minimum at $\tilde{x}_s$. On edges E_3 and E_4, the process can be interpreted as a Brownian motion in a parabolic potential defined on $[\beta, \infty)$ reaching minimum at x_s. A graphic representation of these potentials is shown in Fig. 2.20, with continuous and dashed lines, respectively. The associated transitions probability functions are the solutions of the following Fokker-Planck equations on the corresponding intervals:

$$\frac{\partial}{\partial t}\rho_i(x, t|x_0, 0) = \frac{\partial}{\partial x}\left[b(x - x_s^i)\rho_i(x, t|x_0, 0)\right] + \frac{\sigma^2}{2}\frac{\partial^2}{\partial x^2}\rho_i(x, t|x_0, 0), \qquad (2.90)$$

and subject to the initial boundary conditions described in the previous example, partially given in (2.76). A similar procedure using Laplace transformation can be used to find closed form analytical expressions for these components of the transition probability function in terms of inverse Laplace transforms of the parabolic cylinder functions. Much simpler analytical results can be found for the stationary distributions. The components of the stationary distribution for the OU process defined on graph Z are the solutions of the following equations:

Fig. 2.20 The potential wells for the Brownian motion representing the noise characterization for edges E_1 and E_2 (*continuous line*) and E_3 and E_4 (*dashed line*), respectively

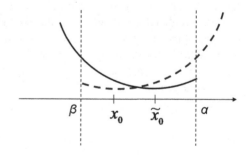

$$\frac{\sigma^2}{2}\frac{\partial^2}{\partial x^2}\rho_i^{st}(x) + \frac{\partial}{\partial x}\left[b(x - x_s^i)\rho_i^{st}(x)\right] = 0, \ldots i = 1, \ldots, 4 \qquad (2.91)$$

and are subject to the following boundary conditions:

$$\rho_1^{st}(\beta^-) = \rho_2^{st}(\beta^+), \ \rho_3^{st}(\beta^+) = 0, \ \rho_3^{st}(\alpha^-) = \rho_4^{st}(\alpha^+), \ \rho_2^{st}(\alpha^-) = 0 \qquad (2.92)$$

$$\frac{\partial \rho_1^{st}}{\partial x}(\beta^-) = \frac{\partial \rho_2^{st}}{\partial x}(\beta^+) + \frac{\partial \rho_3^{st}}{\partial x}(\beta^+), \ \frac{\partial \rho_4^{st}}{\partial x}(\alpha^+) = \frac{\partial \rho_2^{st}}{\partial x}(\alpha^-) + \frac{\partial \rho_3^{st}}{\partial x}(\alpha^-)$$

while $p_1^{st}(x)$ and $p_4^{st}(x)$ have to decay to zero as x goes to minus infinity and plus infinity, respectively.

It is known that the general solutions of linear differential Eqs. (2.91) can be written in the following forms:

$$\rho_i^{st}(x) = e^{-\frac{b(x-\tilde{x}_s)^2}{2\sigma^2}}\left(c_i \int_x^\alpha e^{\frac{b(y-\tilde{x}_s)^2}{2\sigma^2}}dy + d_i\right), \quad i = 1, 2$$

$$\rho_i^{st}(x) = e^{-\frac{b(x-x_s)^2}{2\sigma^2}}\left(c_i \int_\beta^x e^{\frac{b(y-x_s)^2}{2\sigma^2}}dy + d_i\right), \quad i = 3, 4$$

$$(2.93)$$

where c_i and d_i are constants that will be found in our problem from boundary conditions (2.92). The null-conditions $\rho_2^{st}(\alpha^-) = 0$ and $\rho_3^{st}(\beta^+) = 0$ implies $d_2 = d_3 = 0$, while zero decay at minus infinity and plus infinity for $p_1^{st}(x)$ and $p_4^{st}(x)$, respectively, implies $c_1 = c_4 = 0$. Moreover, $\rho_1^{st}(\beta^-) = \rho_2^{st}(\beta^+)$ implies $d_1 = c_2 \int_\beta^\alpha \exp(b(y-\tilde{x}_s)/2\sigma^2)dy$, while $\rho_3^{st}(\alpha^-) = \rho_4^{st}(\alpha^+)$ leads to the relation $d_4 = c_3 \int_\beta^\alpha \exp(b(y-x_s)/2\sigma^2)dy$. The boundary conditions for the derivatives in (2.92) implies $c_2 = c_3$ that will be denoted by c. As a result,

$$\rho_1^{st}(x) = c\left(\int_\beta^\alpha e^{\frac{b(y-\tilde{x}_s)^2}{2\sigma^2}}dy\right)e^{-\frac{b(x-\tilde{x}_s)^2}{2\sigma^2}}, \quad \rho_2^{st}(x) = c\left(\int_x^\alpha e^{\frac{b(y-\tilde{x}_s)^2}{2\sigma^2}}dy\right)e^{-\frac{b(x-\tilde{x}_s)^2}{2\sigma^2}},$$

$$(2.94)$$

$$\rho_3^{st}(x) = c\left(\int_\beta^x e^{\frac{b(y-x_s)^2}{2\sigma^2}}dy\right)e^{-\frac{b(x-x_s)^2}{2\sigma^2}}, \quad \rho_4^{st}(x) = c\left(\int_\beta^\alpha e^{\frac{b(y-x_s)^2}{2\sigma^2}}dy\right)e^{-\frac{b(x-x_s)^2}{2\sigma^2}},$$

where c is determined from the normalization condition for the total stationary probability function $\int_{-\infty}^\beta \rho_1^{st}(x)dx + \int_\beta^\alpha \rho_2^{st}(x)dx + \int_\beta^\alpha \rho_3^{st}(x)dx + \int_\alpha^\infty \rho_4^{st}(x)dx = 1$. An example of the stationary distribution (2.94) obtained for a noise input characterized by $b = 1$, $\sigma = 1$, $x_s = -0.5$, $\tilde{x}_s = 0.5$, and vertex coordinates $\beta = -1$ and $\alpha = 1$ is plotted in Fig. 2.21.

Fig. 2.21 Stationary
probability components for a
"two-wells" Ornstein-
Uhlenbeck process defined on
graph Z

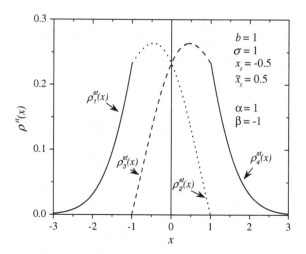

This example of OU process governed by different equations on each edge of the graph is used in describing the stochastic behavior of bistable hysteretic systems where noise is state dependent. In Chap. 5 we will prove that coherence resonance phenomena take place in such system driven by state dependent noise.

References

1. Busch-Vishniac, I. J., West, J. E., Barnhill, C., Hunter, T., Orellana, D., & Chivukula, R. (2005). Noise levels in Johns Hopkins Hospital. *Journal of the Acoustical Society of America, 118*(6), 3629–3645.
2. Vaseghi, S. V. (2008). *Advanced digital signal processing and noise reduction.* New York: Wiley.
3. Perez, R. (1998). *Wireless communications design handbook, volume 3: Interference into circuits: Aspects of noise, interference, and environmental concerns.* San Diego: Academic Press.
4. Néel, L. (1949). Théorie du traînage magnétique des ferromagnétiques en grains fins avec applications aux terres cuites. *Annals of Géophysics, 5,* 99–136.
5. Wernsdorfer, W. (2001). Classical and quantum magnetization reversal studies in nanometer-sized particles and clusters. *Advances in Chemical Physics, 118,* 99.
6. Coffey, W. T., Crothers, D. S. F., Dormann, J. L., et al. (1998). Thermally activated relaxation time of a single domain ferromagnetic particle subjected to a uniform field at an oblique angle to the easy axis: Comparison with experimental observations. *Physical Review Letters, 80,* 5655.
7. Weller şi D., & Moser, A. (1999). Thermal effect limits in ultrahigh-density magnetic recording. *IEEE Transactions on Magnetics, 35,* 4423.
8. Rao, G. N., Yao, Y. D., & Chen, J. W. (2005). Superparamagnetic behavior of antiferromagnetic CuO nanoparticles. *IEEE Transactions on Magnetics, 41,* 3409.
9. Matsumoto, K., Inomata, A., & Hasegawa, S. (2006). Thermally assisted magnetic recording. *FUJITSU Scientific & Technical Journal, 42,* 158.
10. Alex, M., Tselikov, A., McDaniel, T., et al. (2001). Characteristics of thermally assisted magnetic recording. *IEEE Transactions on Magnetics, 37,* 1244–1249.

11. Kosko, B. (2006). *Noise*. London: Viking/Penguin.
12. Ando, B., & Graziani, S. (2000). *Stochastic Resonance*. Dordrecht: Kluwer Academic Publisher.
13. Lindner, B., Garcia-Ojalvo, J., Neiman, A., & Schimansky-Geier, L. (2004). Effects of noise in excitable systems. *Physics Reports, 392*, 321–424.
14. Sagues, F., Sancho, J. M., & Garcia-Ojalvo, J. (2007). Spatiotemporal order out of noise. *Reviews of Modern Physics, 79*, 829.
15. Gammaitoni, L., Hänggi, P., Jung, P., & Marchesoni, F. (2009). Stochastic resonance: A remarkable idea that changed our perception of noise. *European Physical Journal B, 69*, 1–3.
16. Gardiner, C. W. (1997). *Handbook of Stochastic Methods*. Berlin: Springer.
17. Doob, J. L. (1953). *Stochastic Processes*. New York: Wiley.
18. Arnold, L. (1972). *Stochastic Differential Equations*. New York: Wiley.
19. Papoulis, A. (2002). *Probability random variables and stochastic processes*. New York: McGraw Hill.
20. Gihman, I. I., & Skorohod, A. V. (1972). *Stochastic Differential Equations*. Berlin: Springer.
21. Gillespie, D. T. (1996). Exact numerical simulation of the Ornstein-Uhlenbeck process and its integral. *Physical Review E, 54*, 2084–2091.
22. Moss, F., & McClintock, P. V. (1989). *Noise in nonlinear dynamical systems, vol. 1: Theory of continuous Fokker-Planck systems*. Cambridge: Cambridge University Press.
23. Johnson, J. B. (1925). The Schottky effect in low frequency circuits. *Physical Review, 26*, 71–85.
24. Schottky, W. (1926). Small-shot effect and Flicker effect. *Physical Review, 28*, 74–103.
25. Milotti, E. (2002). 1/f noise: A pedagogical review. arxiv preprint: physics/0204033.
26. Ward, L., & Greenwood, P. (2007). 1/f noise. *Scholarpedia, 2*, 1537.
27. http://www.nslij-genetics.org/wli/1fnoise/.
28. Dutta, P., & Horn, P. M. (1981). Low-frequency fluctuations in solids: 1/f noise. *Reviews of Modern Physics, 53*, 497–516.
29. Mandelbrot, B. (1977). *Fractals: Form, chance and dimension*. San Francisco: W. H. Freeman and Co.
30. Bak, P., Tang, C., & Wiesenfeld, K. (1987). Self-organized criticality: An explanation of the 1/f noise. *Physical Review Letters, 59*, 381–384.
31. Bak, P., Tang, C., & Wiesenfeld, K. (1988). Self-organized criticality. *Physical Review A, 221*, 364–374.
32. Yellott, J. I. (1983). Spectral consequences of photoreceptor sampling in the rhesus retina. *Science, 221*, 382–385.
33. Freidlin, M. I. (1996). *Markov processes and differential equations: Asymptotic problems*. Berlin: Springer.
34. Dynkin, E. B. (1965). *Markov processes*. Berlin: Springer.
35. Mandl, P. (1968). *Analytical treatment of one-dimensional markov processes*. Berlin: Academia.
36. Freidlin, M. I., & Wentzell, A. D. (1993). Diffusion processes on graphs and the averaging principle. *Annals of Probability, 21*(4), 2215–2245.
37. Pfeiffer, R. M. (1998). Statistical problems for stochastic processes with hysteresis. Ph.D. Thesis, University of Maryland, College Park.
38. Freidlin, M. I., Mayergoyz, I. D., & Pfeiffer, R. (2000). Noise in hysteretic systems and stochastic processes on graphs. *Physical Review E, 62*, 1850–1856.
39. Mayergoyz, I., & Dimian, M. (2003). Analysis of spectral noise density of hysteretic systems driven by stochastic processes. *Journal of Applied Physics, 93*(10), 6826–6828.
40. Dimian, M., & Mayergoyz, I. D. (2004). Spectral density analysis of nonlinear hysteretic systems. *Physical Review E, 70*, Article 046124.
41. Dimian, M. (2008). Extracting energy from noise: Noise benefits in hysteretic systems. *NANO: Brief reviews and reports, 3*(5), 391–397.
42. Abramowitz M., & Stegun I. (Eds.). (1972). *Handbook of mathematical functions*. New York: Dover Publications.

Chapter 3
Stochastically Driven Hysteretic Systems in Science and Engineering

3.1 Magnetic Hysteresis

The rectangular loop, one of the simplest hysteretic system, can be physically interpreted as the response of the magnetization in a ferromagnetic particle with uniaxial anisotropy under the action of an external magnetic field applied along the anisotropy axis. The particle size is assumed to be small enough to determine a uniform magnetization inside the particle for any values of applied field. The free energy g of such a physical system is outlined in Fig. 3.1 for four representative cases that correspond to points A, B, C and D on the rectangular loop represented in Fig. 3.2. Magnetization values corresponding to a magnetic field are given by one of the minimum points for the free energy. For magnetic fields $h < \beta$ or $h > \alpha$, there exists only one minimum for free energy, namely $m = -1$ (case A) or $m = 1$ (case C). If, however, the magnetic field h is in interval (β, α) then the corresponding free energy has two minimum points $m = -1$ and $m = 1$ (cases B and D).

The particle magnetization may persist in a metastable state for some time, but thermal noise usually drives it to the other metastable state. In specific cases, this transition time is much greater than the observation time and, therefore, the stochastic aspects are neglected. There are situations, however, when the hysteretic system reliability depends on the transition time. In the context of magnetic data storage technology [1], two states ($+1$ and -1) are possible in the absence of magnetic field and they correspond to the values 1 and 0 in the terminology of binary information. By applying an appropriate magnetic field pulse, the magnetization can be switched from one state to another, which means that the memory cell is rewritten. When the transition happens in the absence of the magnetic field due to thermal noise, the stored information is lost. This phenomenon, known as super-paramagnetic effect, is increasingly pronounced as the memory cell gets smaller, being the main obstacle in improving data storage density in magnetic hard disk drives [2–4].

The basic description of superparamagnetic phenomenon is usually provided by using the Brownian motion in a double well potential analyzed in Sect. 2.1.6.

M. Dimian and P. Andrei, *Noise-Driven Phenomena in Hysteretic Systems*, Signals and Communication Technology 218, DOI: 10.1007/978-1-4614-1374-5_3, © Springer Science+Business Media New York 2014

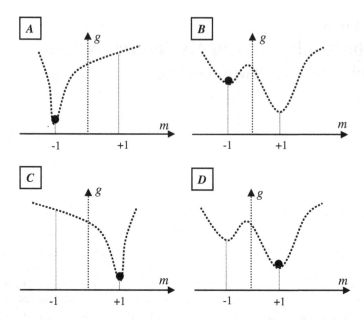

Fig. 3.1 Free energy profile for a magnetic particle with uniaxial anisotropy under several magnetic fields applied along the anisotropy axis

Fig. 3.2 The diagram of the magnetization versus applied magnetic field in a ferromagnetic object with uniaxial anisotropy subject to an external magnetic field applied along the anisotropy axis

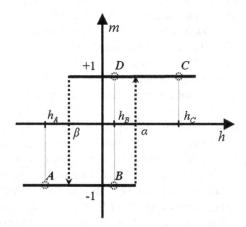

The central result of that section is the Arrhenius formula, also known in magnetism as Néel-Arrhenius formula [5], expressing the mean time of switching between two magnetization states as exponential function of the energy barrier divided by the noise energy. Since the energy barrier is proportional to the volume of the magnetic particle, the mean time of the magnetization switching induced by thermal noise decreases exponentially with the decrease in the particle volume; this fact leads to a minimum volume that the particle needs to have in order to satisfy the stability condition for the recording data. Similarly, since the thermal noise energy is proportional to the absolute temperature, the mean time of the magnetization switching induced by thermal noise decreases significantly with the increase in temperature; this leads to a limit for the operating temperature above which the stability condition for the recording data is not satisfied [6, 7]. These results predicted by Néel-Arrhenius theory have been explicitly proved experimentally by Wernsdorfer and his collaborators [8, 9] in the case of Cobalt nanoparticles.

Next, let us consider a magnetic particle with uniaxial anisotropy subject to a magnetic field applied on a direction non-collinear to the anisotropy axis, denoted by x. Let us consider y-axis in the plane formed by anisotropy axis and the applied field (see Fig. 3.3). It is apparent from symmetry considerations that the particle magnetization lies in the same plane. The part of free energy g dependent on magnetization particle is given by:

$$g = k_a \sin^2 \theta - \mathbf{m} \cdot \mathbf{h}, \tag{3.1}$$

where k_a is the anisotropy constant and θ is the angle between the magnetization $\mathbf{m}$ and the easy axis. The formula is normalized such that the magnetization magnitude is equal to unity. The first energy term is known as anisotropy energy and originates from the intrinsic properties of the particle, such as magnetcrystalline anisotropy or shape anisotropy, while the second energy term is known as Zeeman energy and characterizes the particle interaction with the applied magnetic field.

Since the applied magnetic field is fixed and the magnetization magnitude is assumed constant (and normalized to 1) throughout the process, the free energy is a one-variable function having the following explicit expression:

$$g(\theta) = k_a \sin^2 \theta - h_x \cos \theta - h_y \sin \theta. \tag{3.2}$$

The metastable states of this system correspond to the minima of free energy which can be found among the zeros of its derivatives. Thus, the metastable orientations of the magnetizations are solutions of the following equation:

$$k_a \sin \theta \cos \theta + h_x \sin \theta - h_y \cos \theta = 0 \tag{3.3}$$

Fig. 3.3 Configuration of the magnetization in a small particle with uniaxial anisotropy subject to a uniform magnetic field

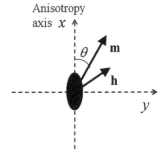

By expressing $\sin \theta$ and $\cos \theta$ from formula (3.3) in terms of $\tan(\theta/2)$, the problem is reduced to a quartic equation with respect to $\tan(\theta/2)$ which may have two or four real solutions, depending on the applied magnetic field. The minimum points can be identified by checking the positivity condition on the second derivative of free energy. Since only one of them can be minimum in the first case and only two in the second case, it can be concluded that (h_x, h_y)-plane is divided in two different regions where the system has one and two metastable states, respectively. On the boundary between these two regions the second derivative of the free energy with respect to θ should also be zero and consequently, these critical magnetic fields (h_x^c, h_y^c) are subject to the following conditions:

$$\begin{cases} 2k_a \sin \theta \cos \theta + h_x^c \sin \theta - h_y^c \cos \theta = 0 \\ 2k_a (\cos^2 \theta - \sin^2 \theta) - h_x^c \cos \theta + h_y^c \sin \theta = 0 \end{cases} \tag{3.4}$$

This linear system with respect to (h_x^c, h_y^c) leads to the following solutions:

$$h_x^c = -2k_a \cos^3 \theta \quad \text{and} \quad h_y^c = 2k_a \sin^3 \theta. \tag{3.5}$$

As a result, the implicit form of the critical curve separating the two regions in (h_x, h_y)-plane is given by:

$$\left(h_x^c\right)^{2/3} + \left(h_y^c\right)^{2/3} = (2k_a)^{2/3}, \tag{3.6}$$

which represents an astroid (see Fig. 3.4) known in magnetic recording community as the Stoner-Wohlfarth (SW) asteroid [10].

In conclusion, when the applied field corresponds to a point in the (h_x, h_y)-plane located inside the SW astroid, the magnetic system features two metastable states, its current state depending on the magnetic field history. When the applied field corresponds to a point located outside the SW astroid, the system features only one metastable state (the equilibrium).

Fig. 3.4 Normalized Stoner-Wohlfarth astroid in the (h_x, h_y)-plane

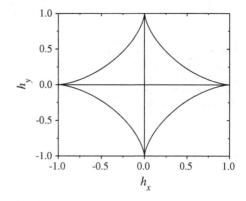

Fig. 3.5 Temperature
dependence of the switching
field of a 3 nm Co cluster,
measured in the plane defined
by the easy and medium hard
axes. Reprint with permission
from [8]

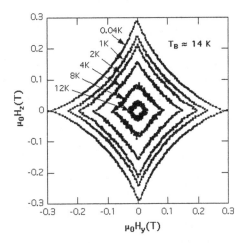

The thermal noise is expected to have a shrinking effect on the critical field curve, fact confirmed experimentally by Wernsdorfer and his collaborators [8] and represented in Fig. 3.5.

In order to study the general effects of thermal noise in such non-symmetric cases, i.e. magnetic field is applied on a direction non-collinear to the anisotropy axis, one needs to take into account the dynamics of the magnetization. The motion of magnetic moment **M** in a uniform magnetic field **H,** can be intuitively decomposed into two components: a precessional motion about the direction of the magnetic field, see Fig. 3.6a, and a relaxation motion that tends to orient the magnetic moment along the direction of the field, see Fig. 3.6b.

The short introduction to modeling magnetization dynamics presented here follows the lines of publication [11] which can be consulted for additional details. First, let us consider the precessional motion, which can be mathematically expressed in the following form:

$$\frac{d\mathbf{M}}{dt} = -\gamma \mathbf{M} \times \mathbf{H}, \tag{3.7}$$

where $\gamma = \mu_0 \gamma_0$, with $\mu_0 = 4\pi 10^{-7} [H/m]$ is the permeability of the vacuum and γ_0 is a quantity characteristic to the magnetic moment that can be determined from experiments or computed by considering the physical origin of the magnetic

Fig. 3.6 Schematic
representation of the
magnetization motion
induced by a magnetic field:
a Precessional component,
b Relaxation effect due to the
damping

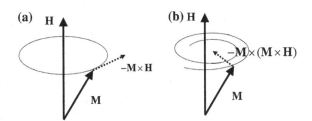

moment. Thus, for the magnetic moment originating from the orbital motion of a particle with mass m_0 and charge q_0 the absolute value of gyromagnetic ratio is $\gamma_0 = g|q_0|/2m_0$, where g is the gyromagnetic factor ($g \sim 2$ for a free electron).

The equation for the precessional motion can be theoretically justified either on quantum or classical grounds. By considering a "static" particle with spin in a uniform magnetic field, the time evolution for the mean value of the spin operator can be derived using either the Schrödinger equation (see [12]) or equivalently, the Von Neumann equation (see [13]). On the classical ground, Eq. (3.7) can be justified assuming that the magnetic moment arises from a "circular" motion of an electron and using Newton's law for the angular momentum, as well as the relation between the magnetic moment and the angular momentum (see [14]). The idea of this classical explanation was first given by Sir Joseph Larmor, and that is why the precessional motion of the magnetic moment about the magnetic field is often called Larmor precession. One may also think of deriving Eq. (3.7) from a variational principle. The magnetic moment $\mathbf{M}$ can be regarded as a "classical top" with principal moments of inertia (0, 0, C), and the Lagrangian formulation of this top case leads to Eq. (3.7) (see [15]). However, each one of these derivations contains assumptions (indicated by quotation marks), which are physically inconsistent. Consequently, they cannot be considered as derivations of Eq. (3.7), but rather as justifications of this equation. In spite of these difficulties, Larmor precession offers accurate explanations of numerous experiments involving the magnetic moment behavior in a uniform magnetic field, such as nuclear magnetic resonance, paramagnetic and ferromagnetic resonance [14], and magnetization reversals [16, 17].

Next, we consider the relaxation motion of the magnetic moment, which can be quantitatively described by adding a dissipative (damping) correction term to Eq. (3.7). The "dissipated" energy is actually transformed by various mechanisms into the thermal energy of the system. Although these mechanisms are partially known [18, 19], they are too complex to be taken into account in an explicit derivation of the damping correction term at a macroscopic level. In order to describe the experimental results, various phenomenological expressions are employed. Most notable ones were given by Landau and Lifshitz for the description of energy losses in the magnetic domain wall motion in ferromagnetic materials [20], and by Bloch for the description of nuclear magnetic relaxation [21]. Since then, these expressions have been successfully applied to various physical phenomena involving dissipation of the magnetic energy. The Landau-Lifshitz expression is mostly used for the description of various dissipative processes in which the norm of the magnetic moment is conserved, while the Bloch expression is appropriate to complementary cases.

The Landau-Lifshitz expression for the damping term reads:

$$-\frac{\gamma \alpha_L}{M^2} \mathbf{M} \times (\mathbf{M} \times \mathbf{H}), \tag{3.8}$$

where α_L is a damping parameter with the dimensionality of the magnetic field, and M is the norm of the magnetic moment $\mathbf{M}$. Nowadays, it is preferred to call dimensionless parameter $\alpha = \alpha_L/M$ as the damping parameter, and this convention is used throughout this book. This damping parameter may take a large interval of values depending on the various types of magnetic materials and experiments involved. The experiments involving magnetization reversals in ferromagnetic thin films and nanoparticles used in magnetic data storage indicate rather small values for the damping parameter on the order of 10^{-3} up to 10^{-1}.

Bloch used the following expression for the damping term:

$$\left(-\frac{M_x}{\tau_2}, -\frac{M_y}{\tau_2}, \frac{M_s - M_z}{\tau_1}\right), \tag{3.9}$$

where z is the direction of the applied magnetic field, M_s is the magnetization saturation value in the equilibrium state, and the parameters τ_1 and τ_2 account for the relaxation times in the longitudinal and transverse directions, respectively. Since we focus here on magnetization dynamics that conserves the norm of the magnetic moment, we do not extend further the discussion of Bloch equation.

The dissipative term given by formula (3.8) is somehow atypical, in the sense that it cannot be derived in terms of the standard Rayleigh dissipation function. By using the standard Rayleigh function to introduce the dissipative effects in the Lagrangian formulation for the conservative precessional motion, Gilbert [22] derived the following damping term:

$$-\frac{\gamma \alpha_G}{M} \mathbf{M} \times \frac{d\mathbf{M}}{dt}, \tag{3.10}$$

where $\alpha_G > 0$ denotes the Gilbert dimensionless damping parameter.

In conclusion, the general equation of motion for a magnetic moment in homogeneous applied field can be written as:

$$\frac{d\mathbf{M}}{dt} = -\gamma_G \mathbf{M} \times \mathbf{H} - \frac{\gamma_G \alpha_G}{M} \mathbf{M} \times \frac{d\mathbf{M}}{dt}, \tag{3.11}$$

in the Gilbert form, known as Landau-Lifshitz-Gilbert equation. Here, γ_G is identical to γ, but the G index is used for the clarity of future considerations. By using the damping term given by formula (3.8), the equation of motion has the following form:

$$\frac{d\mathbf{M}}{dt} = -\gamma \mathbf{M} \times \mathbf{H} - \frac{\gamma \alpha}{M} \mathbf{M} \times (\mathbf{M} \times \mathbf{H}). \tag{3.12}$$

This equation is known as Landau-Lifshitz equation. It can be simply proved that Eqs. (3.11) and (3.12) are mathematically equivalent. However, there is some physical discrepancy between them if one thinks at γ as a physical constant with its value given by the physical origin of the magnetic moment. Nevertheless, this discrepancy is considerable diminished when α is a small parameter, as it is usually the case in practical applications.

The Landau-Lifshitz equation can be easily generalized to describe the behavior of the magnetic moment in complex environments substituting the magnetic field $\mathbf{H}$ by the effective field $\mathbf{H}_{eff}$, which is derived using the magnetic energy

g associated to the environment $\mathbf{H}_{eff}(\mathbf{M}) = -\nabla_{\mathbf{M}}g$. By normalizing the magnetization $\mathbf{m} = \mathbf{M}/M$, and the effective field $\mathbf{h}_{eff} = \mathbf{H}_{eff}/M$, the Landau-Lifshitz equation can be written in the following dimensionless form:

$$\frac{d\mathbf{m}}{dt} = -\mathbf{m} \times \mathbf{h}_{eff} - \alpha \, \mathbf{m} \times \mathbf{m} \times \mathbf{h}_{eff}, \qquad (3.13)$$

where time is measured in units of $(\gamma M)^{-1}$.

The thermal noise effects can be included in this description by considering a vector white noise in the effective magnetic field:

$$\mathbf{h}_{eff}(\mathbf{m}) = -\nabla_{\mathbf{m}}g + \mathbf{h}^{th}(t) \qquad (3.14)$$

The thermal field is explicitly defined by zero-mean normal variables with correlations:

$$<h_i^{th}(t_1)h_j^{th}(t_2)> \; = 2\alpha \frac{kT}{v}\delta_{ij}\delta(t_1 - t_2) = 2\frac{\alpha}{\beta}\delta_{ij}\delta(t_1 - t_2) \qquad (3.15)$$

where indices $i, j = 1, 2, 3$ correspond to the Cartesian coordinates x, y, z, k is the Boltzman's constant, T is the absolute temperature and v is the particle volume. Actually, according to the fluctuation–dissipation theorem [23] the introduction dissipation term requires a fluctuating terms. The Fokker–Planck equation for the probability density function ρ of the magnetization was first derived by using Wang and Uhlenbeck methods [24]. A simpler derivation has been later provided by Coffey in [25] based on Einstein's approach. This so-called Brown's Fokker–Planck equation has the following form:

$$(1 + \alpha^2)\sin\theta \frac{\partial\rho}{\partial t} = \frac{\partial}{\partial\theta}\left\{\sin\theta\left[\left(\alpha\frac{\partial g}{\partial\theta} - \frac{1}{\sin\theta}\frac{\partial g}{\partial\varphi}\right)\rho + \frac{\alpha}{\beta}\frac{\partial\rho}{\partial\theta}\right]\right\}$$
$$+ \frac{\partial}{\partial\varphi}\left\{\left(\frac{\partial g}{\partial\theta} - \frac{\alpha}{\sin\theta}\frac{\partial g}{\partial\varphi}\right)\rho + \frac{\alpha}{\beta\sin\theta}\frac{\partial\rho}{\partial\varphi}\right\} \qquad (3.16)$$

In the particular case of uniaxial anisotropy with the magnetic field applied along easy axis, the free energy g is independent of φ and can be reduced to the following expression:

$$g(\theta) = -\frac{1}{2}\cos^2\theta - h_x\cos\theta = -\frac{1}{2}m_x^2 - h_xm_x \qquad (3.17)$$

Consequently, the probability density function is also independent of φ and its expression in terms of m_x will be denoted by $p(m_x, t)$ leading to a much simpler expression for Brown's Fokker–Planck Eq. (3.16):

$$\frac{1 + \alpha^2}{\alpha}\frac{\partial p}{\partial t} = \frac{\partial}{\partial m_x}\left\{(1 - m_x^2)\left[\frac{dg}{dm_x}p + \frac{1}{\beta}\frac{\partial p}{\partial m_x}\right]\right\} \qquad (3.18)$$

The stationary equation obtained by making the left hand side equal zero, can be simply integrated once with respect to m_x, so the stationary probability function p_s satisfies the following relation:

$$(1 - m_x^2)\left[\frac{dg}{dm_x}p_s + \frac{1}{\beta}\frac{dp_s}{dm_x}\right] = 0 \tag{3.19}$$

where the constant of integration was found to be zero by evaluating the expression from the left hand side at $m_x = 1$. It is apparent from (3.19) that the factor inside the square parentheses must be zero, which can be rearranged as follows:

$$\frac{\partial p_s(m_x)}{\partial m_x} = -\beta \frac{\partial g(m_x)}{\partial m_x} p_s(m_x) \tag{3.20}$$

which can be easily solved by separation of variables. In conclusion, the stationary distribution for the magnetization motion of a magnetic particle with uniaxial anisotropy driven by an external magnetic field h applied along the anisotropy axis and by thermal noise of energy $1/\beta$ is found:

$$p_s(m_x) = a_0 e^{\beta((1/2)m_x^2 + h_x m_x)} \tag{3.21}$$

where a_0 is found from the normalization condition for p_s. The non-stationary distribution, *i.e.* the solution of Eq. (3.18), can be expressed as a series:

$$p(m_x, t) = p_s(m_x) + \sum_{n=1}^{\infty} a_n p_n(m_x) e^{-\lambda_n t} \tag{3.22}$$

where $p_n(m_x)$ and λ_n are the eigenfunctions and eigenvalues, respectively, associated to the Eq. (3.18). Thus, they are the solutions of the following Sturm–Liouville problem on $[-1, 1]$:

$$\frac{d}{dm_x}\left\{(1 - m_x^2)e^{-\beta g(m_x)}\frac{d}{dm_x}\left(e^{\beta g(m_x)}p_n(m_x)\right)\right\} - \frac{\beta(1 + \alpha^2)}{\alpha}\lambda_n p_n(m_x) = 0 \tag{3.23}$$

Most of the articles published in this area in 1960's and 1970's have presented analytical formulas and numerical results for the eigenvalues and eigenfunctions in this axially symmetric case. Except in the early stages of an approach to equilibrium, the only appreciably time dependent term in the series solution (3.22) is $p_1(m_x)$, which corresponds to the longest relaxation time $(1/\lambda_1)$.

In the general case, the magnetic field can be applied at any angle with respect to the anisotropy axis, while the anisotropy may have non-uniaxial characteristics. Consequently, the symmetry of the problem breaks and the solution of the Eq. (3.16) should be found in the general form:

$$\rho(\theta, \varphi, t) = \rho_s(\theta, \varphi) + \sum_{n=1}^{\infty} \tilde{a}_n \rho_n(\theta, \varphi) e^{-\tilde{\lambda}_n t} \tag{3.24}$$

where $\rho_n(\theta,\varphi)$ and $\tilde{\lambda}_n$ are the eigenfunctions and eigenvalues, respectively, associated to the Eq. (3.16). The analysis of this equation is beyond the scope of this book but the reader interested in this topic can consult the excellent monograph published by Coffey, Kalmykov and Waldron [25]. Here, we just remind the result presented in Fig. 3.5 about the shrinking effect of the thermal noise on the critical field curve confirmed experimentally by Wernsdorfer and his collaborators in the case of uniaxal anisotropy [8].

Most of the magnetic materials have much more complex behavior and can not be framed into the uniform magnetization approach presented above. As an example, Fig. 3.7 illustrates the nonuniform magnetic structure of a micrometric part from a magnetic recording media (a film of cobalt with iron-oxide doping). A NT-MDT atomic force microscope is used in magnetic force microsopy mode to visualize the topography and magnetic profile of the given micrometric sample. In order to address this non-uniform behavior, multi-spin and micromagnetic models are involved at microscale and hysteretic models are used at macroscale.

A common discrete approach to describe the magnetization behavior in nano-structures is based on Landau-Lifshitz-type equation for each magnetic moment with the effective field obtained from a Heisenberg-type Hamiltonian including the external applied field, exchange interaction, dipole–dipole interaction, and the contribution of magnetocrystalline anisotropy. This semi-clsssical multi-spin description of magnetic structures leads to complex many-body problems, which are normally tractable only by numerical methods. In Fig. 3.8 are shown samples of metastable magnetic configurations obtained using this approach for a spherical nanoparticle featuring a radial anisotropy for the surface spins and uniaxial anisotropy for core spins. Due to computer limitations, the investigations are restricted to very small systems with diameters of a few nanometers. Therefore, the only way to approach larger magnetic systems is to ignore the discrete nature of matter and to use a continuum approximation.

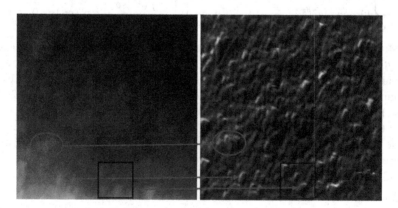

Fig. 3.7 Topography (*left*) and magnetic profile (*right*) of a magnetic recording media obtained by using a magnetic force microscope

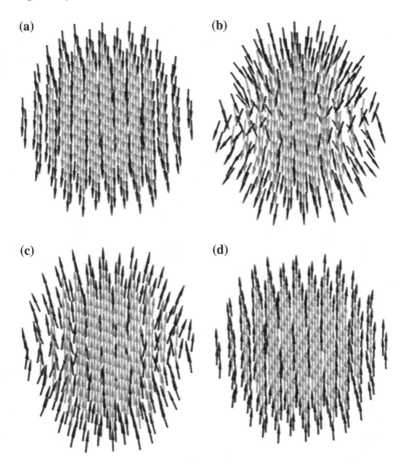

Fig. 3.8 Samples of magnetic configuration for a multi-spin system corresponding to saturation states (negative (**a**) and positive (**d**)) and two intermediate states (negative (**b**) and positive (**c**)). *Gray arrows* represent core spins and *black arrows* represent surface spins

A popular continuum-mechanics approach to describe the magnetization behavior can be traced back to the works of Landau and Lifshitz, Brown, and Aharoni, and it is known as micromagnetics. A general presentation of the micromagnetics can be found in the classical book of Brown [15] and in the one edited by Rado and Suhl [26]. A seminal overview of this domain is presented in the book of Aharoni [27], while recent reviews of analytic and numerical micromagnetics, compared to experimental results, can be found in the books of Kronmuller and Fahnle [28], and Hubert and Schafer [29].

Micromagnetics theory has been mainly applied to the calculation of quasistatic magnetization processes in micrometric structures. The classical approach was based on the analytical and numerical study of the static Brown equation and its linearized form. The increasing computational capabilities made possible a new approach based on the dynamic Landau-Lifshitz equation with the effective field derived from the Brown Hamiltonian. However, the complexity of the dynamic approach may generate unacceptable large errors in describing long-time scale processes. For example, a standard problem proposed by the National Institute for Standards and Technology (NIST) was simulated by various computational groups and the numerical results had been submitted during 1997–1998 [30]. The wide distribution of these results raised many doubts concerning the reliability of the numerical methods applied to solve this complex problem. As a consequence, NIST proposed simpler standard problems related to the short time scale processes. In this case, the submitted numerical results tend to agree with each other on a time scale below 1 ns. A large number of research articles and Ph.D. theses concentrate nowadays in this area, providing valuable numerical algorithms to approach Landau-Lifshitz type equations for continuum media.

As an example, this approach is used in Fig. 3.9 to illustrate the magnetic vortices in spherical Permalloy particles the vortex. The simulation is performed by using Nmag software [31], which is a finite-element micromagnetic simulation package based on Python scripts running on a Linux virtual machine. For a better visualization, the magnetization of each finite element of the continuous structure is represented by an arrow. By increasing the applied magnetic field from a large negative value generating the negative saturation of the magnetization (a) to a large positive value generating the positive saturation (f), the magnetization evolves through various vortex states (b)–(e).

In conclusion, micromagnetics offers a valuable tool for describing the magnetization behavior in magnetic materials, but the complexity of this integro-differential problem limits the computational analysis to micrometer structures. This scale limitation is even greater when the stochastic form of the Landau-Lifshitz equation is used to analyze the thermal effects on magnetization behavior subject to applied fields. Numerous phenomenological models have been developed to describe the magnetization behavior at larger scales with a special interest in hysteresis.

By using Preisach formalism, complex hysteretic behavior could be described as a weighted superposition of responses of elementary hysteresis operators to the given magnetic field. Preisach models have been extensively used in modeling hysteretic phenomena in magnetic materials with various applications ranging from magnetic recording technologies to electrical machines. A description of Preisach models with deterministic inputs has been provided in Chap. 1.

Preisach-type models with stochastic input were introduced by Mayergoyz and Korman to offer a unified and detailed description of hysteresis and "after-effect" in magnetic materials [32–35]. Key computations in these viscosity models are based on the relation between randomly induced switchings of rectangular loops and the exit problem for stochastic processes, which is a well studied problem in the theory of diffusion processes. Later, another technique for these computations was discovered which uses the recently developed theory of diffusion processes on graphs [36] developed by Freidlin and Wentzell. This theory was first applied to the study of random perturbations of Hamiltonian dynamical systems [36, 37]. Then it was realized that this mathematical technique is naturally suited to the analysis of noise in hysteretic systems [38–41]. In the following chapters, this

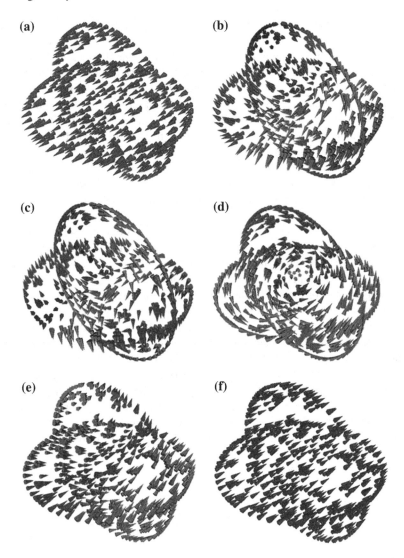

Fig. 3.9 Samples of magnetic configuration in horizontal and vertical plane for a spherical magnetic particles using micromagnetic modeling. By increasing the applied magnetic field, magnetization switches from the negative (**a**) to positive (**f**) saturation state using a vortex mechanism reflected in (**b**)–(**e**)

model will be extensively used for analytical and numerical analysis of thermal relaxation and spectral noise density of hysteretic systems.

Although Preisach models have enjoyed a dominant role in hysteresis modeling of magnetic materials over many years, other hysteresis models such as the Jiles-Atherton, energetic, and Coleman-Hodgdon models have shared its success and popularity.

The Jiles-Atherton model (see Sect. 1.3) is particularly suited for soft magnetic materials, in which the magnetization change is due to domain wall motion [42, 43]. The restraining force on the displacement of the domain walls is caused the pinning of the walls at the defect sites and expressed in terms of the pinning constant, k. The interaction between the magnetic moments is modeled using the mean-field approximation (see Sect. 1.7.1), and captured by a single mean-field parameter α. Subsequent versions of the model have included the reversible change in magnetization in order to reduce the difference between the anhysteretic and irreversible magnetization [44]. This reversibility is captured by parameter c. Together with the saturation of the magnetization and the initial slope of the anhysteretic curve the model can be written as an ordinary differential equation with 5 parameters, which are related to the physical phenomena in the magnetic system.

The energetic model is suited for both crystals and systems of magnetic particles (see Sect. 1.4). The model equations were derived by minimizing the total magnetic energy of the system, which included reversible and irreversible components [45]. The reversible energy consists of magnetocrystalline and shape anisotropy and is responsible for the rotation of the magnetization at strong fields and for the reversible wall motion at low fields. The irreversible energy is caused by pinning centers, such as grain boundaries, nonmagnetic inclusions, and misorientations of the crystallites. By carefully including the grain orientations and the energy of the applied field, the model can describe accurately magnetization curves in all crystallographic directions [46]. Although the derivation of the model equations is somewhat more complicated than in the case of the Jiles-Atherton model, the energetic model can be expressed as a transcendental equation for the magnetization. The 6 parameters of the model are strongly related to the physical phenomena that take place during the magnetization change. Both the Jiles-Atherton and the energetic models have suffered a number of improvements over the years and were adjusted to model stress and temperature dependent magnetizations [47–50].

The Coleman-Hodgdon model (see Sect. 1.6) was developed as a purely mathematical model of hysteresis without offering any insight into the physics of the magnetization process [51, 52]. The model parameters do not relate too much with the physical mechanisms of the magnetization change and they need to be identified carefully by using proper fitting techniques such as least-square minimization or evolutionary techniques (see Sect. 1.1.3). Although somewhat criticized for its non-physical structure, the model was used successfully to describe magnetic hysteresis in superconductors [53], ferrite [54] and ferromagnetic materials [55]. Subsequent modifications of the model allowed the description of rate-dependent hysteresis with remarkable accuracy [54, 56].

The software accompanying this book (HysterSoft©) can perform different types of simulations with deterministic or noisy inputs for any of the above hysteresis model, predefined or user-defined. Numerous examples are provided in the next chapters of this book.

3.2 Hysteresis in Mechanical Systems

The study of hysteretic phenomena in mechanical systems has a similar history to
the one in magnetism. It seems to date back to the work of Ewing [57] who
analyzed the deformation of a steel wire produced by an external force and noticed
that the deformation did not disappear completely when the force was removed.
Moreover, by applying a cycling stress to the material, a closed curve is obtained
in the stress–strain space. When a cycling stress is applied to the material, the
strain describes a closed curve as a function of stress. This phenomenon attributed
to fatigue is called elastic hysteresis; sometimes it is also known as plastic hysteresis being a dual manifestation of the elasto-plasticity characteristic of the
hysteretic material. The systematic analysis of elasto-plasticity was initiated by
Prandtl [58, 59] in the 1920's based on superposition of "stop" operators, which
were presented in Sect. 1.2.6.

Twenty years later similar results were rediscovered by Ishlinski [60] and the
model became popular as Prandtl-Ishlinski models. Fig. 3.10 presents sample
hysteresis loops of the strain as a function of the stress in steel and closely
resemble to the results obtained by Prandtl-Ishlinski model.

Such hysteresis loops have been obtained experimentally and theoretically by a
number of authors and for many types of the elastic material [61–72]. Much of the
work in the literature focuses on the analysis of stress–strain response under different geometries, temperatures, loading conditions, and for systems with more
than one degree of freedom. The hysteretic properties of elastic systems are usually
needed for the estimation of crack growth rates and lifetime of these systems.

Many other mechanical systems present hysteresis. Most often the hysteresis is
due to friction forces that appear between various mechanical components of the
system. In addition to the special case of elasto-plasticity previously discussed, we
analyze in this section we analyze a few examples that are often encountered in
practical applications. First, we look at two simplified mechanical systems based
on elastic springs, after which we summarize a few other applications which
involve the Bouc-Wen model.

Fig. 3.10 Sample stress–
strain hysteresis loops

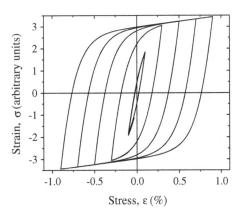

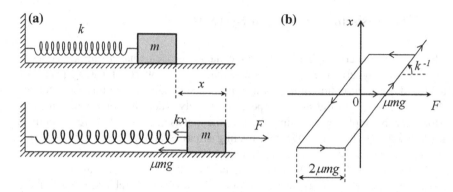

Fig. 3.11 Example of a mechanical system with hysteresis (**a**) in which the displacement x as a function of applied force F can be modeled by the backlash (play) hysteresis operator (**b**)

Let us consider a spring with stiffness k connected to a single body of mass m and to the wall (see Fig. 3.11a). We assume that the spring is initially unstretched and the coefficient of friction between the body and the horizontal surface is μ. If x is the displacement of the body from the initial position when a quasistatic force F is applied to the body, it is straightforward to show that:

$$kx(t) = \max[F(t) - \mu mg, \min(F(t) + \mu mg, kx(t_-))] \tag{3.25}$$

which is nothing else but the backlash operator presented in Sect. 1.2.6 and shown in Fig. 3.11b. The initial force required to move the mass away from the initial position is equal to maximum friction force μmg, where g is the gravitational acceleration. The slope of the oblique lines in Fig. 3.11b is equal to the inverse of the spring stiffness, $1/k$.

As discussed in Chap. 1, the backlash operator is a particular type of the Preisach model, in which the Preisach function is distributed on a line parallel to the $\beta = \alpha$. Hence, the mechanical system presented in Fig. 3.11a can be modeled by the Preisach model in which the distribution in given by (see Fig. 3.12):

$$P(\alpha, \beta) = \frac{1}{k}\delta(\alpha - \beta - \mu mg) \tag{3.26}$$

where δ is the Dirac delta function.

In the above analysis we have assumed that the body is always in mechanical equilibrium and the total force acting on it is zero. If the external force $F(t)$ is not quasi-static and the body can accelerate and decelerate, the dynamics of the system can be modeled by the following equations, which also involve the backlash operator:

$$m\ddot{x}(t) + v\dot{x}(t) + f(t) = F(t) \tag{3.27}$$

$$kx(t) = \max[f(t) - \mu mg, \min(f(t) + \mu mg, kx(t_-))] \tag{3.28}$$

Fig. 3.12 The Preisach plane corresponding to the mechanical system shown in Fig. 3.11a

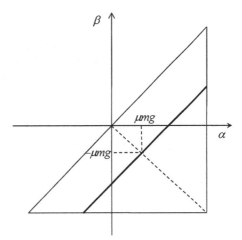

where the overdot denotes the derivative with respect to time and $f(t)$ is the restoring force. To extend the generality of the problem we have assumed in (3.27) that mass m is also subject to a damping force proportional to the velocity $F_{damping} = -v\dot{x}(t)$.

Now let us focus on a more complex mechanical system containing n bodies of masses $m_1, m_2, \dots m_n$ connected to n springs with stifnesses $k_1, k_2, \dots k_n$ like in Fig. 3.13. If we denote by $x_1, x_2, \dots x_n$ the relative displacements of each body with respect to the initial position and we assume that force F acts quasistatically, it can be shown that the displacement of the last body x_n can be modeled by a super-position of backlash operators, which can be represented in the Preisach plane by a distribution of lines along the interaction axis $\beta = \alpha$ (see Fig. 3.14). The coordinates $f_1, f_2, \dots f_n$ of these lines can be expressed as a function of the coefficients of friction and the masses of each body:

$$f_n = \mu_n m_n g, f_{n-1} = f_n + \mu_{n-1} m_{n-1} g, \dots, f_1 = f_2 + \mu_1 m_1 g \qquad (3.29)$$

The Preisach distribution can be mathematically written as:

$$P(\alpha, \beta) = \sum_{k=i}^{n} \frac{1}{k_i} \delta(\alpha - \beta - f_i) \qquad (3.30)$$

where f_i are given by (3.29). It is apparent from this discussion that if the system consists of an infinity of continuously distributed springs and masses, the total displacement can be modeled by the Preisach model, in which the distribution is continuous.

It should be noted that there are many works in the literature that aim to develop hysteresis models based on the superposition of backlash (and stop) operators. One of the most cited such superposition is the Prandtl-Ishlinskii model, which was

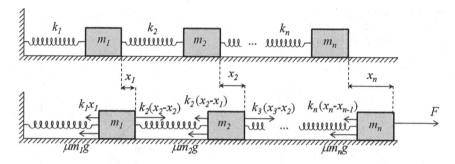

Fig. 3.13 Mechanical system in which the displacement of the last block, x_n, can be modeled by a distributed set of backlash operators

Fig. 3.14 The Preisach plane corresponding to the mechanical system shown in Fig. 3.13

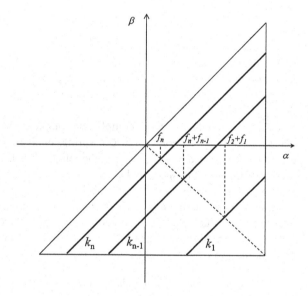

initially proposed by Prandtl [59] and then rediscovered by Ishlinskii [60]. This model can also be written in terms of the Preisach model of hysteresis.

For the remaining of this section we summarize a few other mechanical systems that present hysteresis and have been modeled in the literature by using the Bouc-Wen model. A common feature of all the mechanical systems is that they are governed by the Newton's second law of motion, which can be rearranged to obtain an analytical expression for the restoring force $f(t)$ as function of the excitation force $F(t)$:

$$f(t) = F(t) - m\ddot{x}(t) - v\dot{x}(t) \tag{3.31}$$

where $x(t)$ denotes the relative displacement of the mechanical system. Equation (3.31) has been written for the case when only one mechanical component is

Fig. 3.15 Representation of mechanical systems with hysteresis described by the Bouc-Wen model

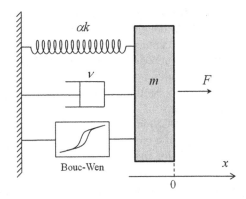

allowed to move in only one direction (i.e. one degree of freedom), however, it can easily be generalized to more mechanical components and 3-dimensional systems. In this case (3.31) becomes a system of $3N$ equations where N is the number of degrees of freedom. The restoring force can be decomposed into an elastic $\alpha k x(t)$ and a hysteretic part $(1 - \alpha)Dkz(t)$, where α, k, and D are some given parameters:

$$f(t) = \alpha k x(t) + (1 - \alpha)Dkz(t). \tag{3.32}$$

In the above equation $z(t)$ is a hysteretic parameter that could be in principle modeled by any hysteretic model. In mechanical systems $z(t)$ is often model by the Bouc-Wen model described in Sect. 1.5.

Model (3.31)–(3.32) is represented schematically in Fig. 3.15. This model has been used in the literature to describe the dynamic performance of seat suspensions in off-road machines [73], non-linear hysteretic absorbers and dampers [74–77], magnetorhelological dampers [78, 79], rubber bushing in vehicle suspension models [80], piezo-electric actuators [81], seismic protection device (such as elastomeric base isolators and buckling restrained dissipative braces) [82], degradation and pinching of structural systems under seismic demand [83], and to analyze random vibrations in mechanical hysteretic systems [84–87].

The accurate modeling of hysteretic systems driven by stochastic input opens opportunities to optimize and design the mechanical systems for controlling unwanted effects such as vibrations, throbbing, fluctuations and noise.

3.3 Hysteresis in Superconductive Materials

It is well-known that thermal noise is affecting the resistivity of a material, which decreases when lowering the temperature. What came as a surprise one century ago was that this monotonic behavior reaches zero at a finite temperature, coined as critical temperature. Onnes first proved that the resistance of mercury abruptly disappears at 4.19 K and an electric current generated by a battery in a superconductor ring persists with the same intensity long after the battery was removed. Although a complete and satisfactory theoretical explanation emerged only half a century later, the breakthrough of the experimental discovery was immediately recognized and brought Onnes the Nobel Prize in 1913. In the subsequent decades, this phenomenon has been observed in several other materials but it seemed to be limited by a critical temperature of 30 K according to some interpretations of the microscopic theory proposed by Bardeen, Cooper and Schrieffer (BCS) in the late 1950's. Consequently, the 1986 discovery by Bednorz and Müller of a cuprate superconductor with a transition temperature of 35 K has opened a major challenge for the theoretical physics. Many other superconductors have been discovered since then, reaching a critical temperature of 138 K at ambient pressure and even 164 K under high pressure [88]. All these high temperature superconductors are characterized by a continuous second order phase transition from superconductive state to normal state within an increasing magnetic field which was explained by Abrikosov (Nobel Prize, 2003) based on the formation of vortex lattices in magnetic fields. It was found that these so-called type-II superconductors exhibit an irreversible behavior in the sense that a change in the monotonicity of an input parameter, such as temperature, pressure, applied electric or magnetic field, does not lead to a reversible change of the system but rather to a hysteretic effect. These effects are well captured by the Bean Model, which resembles the hysteretic models of ferromagnetism but the role of magnetization domains is played in superconductors by the domains of current density [89, 90].

The Bean Model postulates that for weak applied fields or currents, the outer part of the sample has a constant magnitude of the super current density, denoted by J_c, while the interior is shielded from these fields and currents. When the applied field or current increases in magnitude, the zero-region shrinks and vanishes for sufficiently strong applied field or current. In the non-zero region, also known as critical state, the relation between magnetic field $\mathbf{H}$ and current density $\mathbf{J}$ is given by the Maxwell curl equation:

$$\nabla \times \mathbf{H} = \mathbf{J} \qquad (3.33)$$

The analysis presented bellow follows the line of book [91].

Let us examine the case of a rectangular slab oriented as shown in Fig. 3.16 subject to a magnetic flux applied along z-direction. It is natural to assume that the magnetic field inside the slab is also oriented along vertical direction and consequently, the current density is in the horizontal plane.

If we neglect the current density component J_x at the ends of the loops, Eq. (3.33) becomes:

$$\frac{d}{dx} H_z(x) = J_y(x) \qquad (3.34)$$

According to the Bean's postulate on the critical state in the low field case, the current density has the following expression:

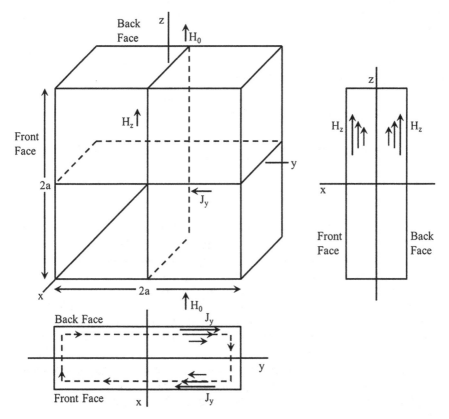

Fig. 3.16 Superconducting slab of thickness 2a subject to an applied magnetic field $\mathbf{H_0}$ directed along z. The induced shielding current density $\mathbf{J}$ flowing in the y direction inside a z-cross-section is presented *below* and total field inside y-cross-section is presented on the *right*

$$J_y(x) = \begin{cases} J_c, & -a \le x \le -a' \\ 0, & -a' \le x \le a' \\ -J_c, & a' \le x \le a \end{cases} \tag{3.35}$$

where a is the width of the slab and a' is the penetration depth of the critical state. By solving Eq. (3.34) under this assumption, one obtains a linear dependence of the internal magnetic field on x in the critical state region, as follows:

$$H_z(x) = \begin{cases} H_0\left(\frac{a'+x}{a'-a}\right), & -a \le x \le -a' \\ 0, & -a' \le x \le a' \\ H_0\left(\frac{x-a'}{a-a'}\right), & a' \le x \le a \end{cases} \tag{3.36}$$

where the internal field is subject to boundary conditions $H_z(a') = 0 = H_z(-a')$ and $H_z(a) = H_0 = H_z(-a)$, where H_0 is the magnitude of the external applied field. As a result, the critical current density J_c and applied magnetic field H_0 are related to each other by the following expression:

$$J_c = \frac{H_0}{a - a'}. \tag{3.37}$$

In the high field case, the critical state extends over the entire sample, so the current density and internal magnetic field read as follows:

$$J_y(x) = \begin{cases} J_c, & -a \le x \le 0 \\ -J_c, & 0 \le x \le a \end{cases} \quad \text{and} \quad H_z(x) = \begin{cases} H_0 - H^* \left(\frac{a+x}{a} \right), & -a \le x \le 0 \\ H_0 + H^* \left(\frac{x-a}{a} \right), & 0 \le x \le a \end{cases}.$$
$$\tag{3.38}$$

where characteristic field H^* is related to the critical current density $H^* = J_c a$.

It is apparent that the increase in the applied surface field H_0 generates the increases in internal field and current density which also proceed inwards. The snapshots presented in Fig. 3.17a show the field configurations for x axis as the applied field H_0 is successively increased from 0 to $2H^*$.

If we consider that the magnetic field is decreasing after achieving the maximum value, the previous profile is partially wiped out by the new linear profile. Such an example is provided in Fig. 3.17b, where the external field is first increased up to $H^*/2$ and then decreased to 0.

Next now let us proceed to the calculation of the average magnetic flux density B since, in practice, B and H_0 are experimentally measured and their relation was found to exhibit hysteresis. The average magnetic flux B_z is defined as follows:

$$B_z = \frac{\mu_0}{2a} \int_{-a}^{a} H_z(x) dx \tag{3.39}$$

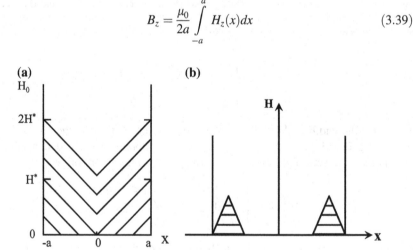

Fig. 3.17 a Internal field in a superconducting slab for increasing values of the applied field from 0 to $2H^*$; **b** Final configuration of internal field when the applied field is first increased up to $H^*/2$ and then decreased to 0

Let us assume that no magnetic field and no current density are present in the superconductor at the initial instant of time t_0.

A back-and-forth variation of the external magnetic field $H_0(t)$ reaching a maximum value H_m $(< H^*)$ is then applied to the sample. The internal field configuration at time t_1 when the external magnetic field is reduced to zero is similar to the one shown in Fig. 3.17b and, by using definition (3.39), the following expression of the average magnetic flux is found:

$$B_z(t_1) = \frac{\mu_0 H_m^2}{4H^*} \tag{3.40}$$

This clearly suggests the hysteretic behavior exhibited by B versus H_0 relation in this superconductor model. In general, the ascending and descending branches of the hysteresis loop obtained by magnetic field oscillation between H_m and $-H_m$ can be described by the following formula:

$$B_z(t) = \pm\mu_0 \left[\frac{H_m^2}{2H^*} - \frac{(H_m \mp H_0(t))^2}{4H^*} \right] \tag{3.41}$$

where the upper signs correspond to the descending branch of the loop, while the lower signs correspond to the ascending branch. Similar procedures can be employed to derive B versus H_0 relation for other profiles of the external magnetic field.

This Bean model captures the main features of the hysteretic behavior in superconductor and has been used by many experimentalists for data interpretation and characterization. It was also subject to various generalizations (known as critical state models) which considered different relationships between the internal field and current density. Mayergoyz and his collaborators have proven that wiping-out property and congruency property hold for Bean model and consequently, it can be described by the Preisach model presented in Chap. 1 [92, 93].

At high enough temperature, a decay of magnetic flux is observed in most of type II superconductors. This phenomenon, known as flux creep, may be interpreted as a stochastic change of flux profile do to the thermally activated jumps of magnetic vortices from one pinning state to another. The behavior can be mathematically described by employing hysteretic models driven by stochastic input, similar to the approach of after-effects in magnetic materials discussed in the first section of this chapter.

3.4 Hysteresis in Molecular Materials

The remarkable evolution in miniaturization process of technological devices is rapidly approaching the molecular scale, and consequently, it has led to an intensive research effort on molecular materials. They are attractive not only for their miniaturization potential but also for novel properties and enhanced performance achieved by using synthetic chemistry. The challenge remains to master, at the molecular level, the traditional functions of information storage, communication and processing.

An important class of molecular materials is represented by spin-crossover compounds (SCO) that have become of great interest recently due to their potential applications in memories, sensors, switches, and display devices [94–96]. These materials are particularly interesting because upon application of heat, light, pressure or a magnetic field they feature a phase transition between a low-spin (LS) diamagnetic ground state and a high-spin (HS) paramagnetic state, accompanied by color and volume change. This short introduction to SCO follows the lines of publications [97, 98] which can be consulted for additional details.

Spin crossover phenomena occurs in some octahedrally coordinated transition metal compounds with 4–7 electrons in the 3d orbital. As a consequence of the splitting of the energy of d orbitals into the t_{2g} and e_g sets in a ligand field, octahedral complexes may exist in high or low spin states. The transition takes place due to the competition between δ and Π, where δ is the splitting energy between the t_{2g} and e_g orbitals and Π is the spin pairing energy (see Fig. 3.18). The spin crossover was first observed experimentally by Cambi and Szego [99] in 1931 when studying temperature variation of the magnetic susceptibility in a series of Fe(III) $3d^5$ compounds. Nowadays, a large and expanding SCO compounds database is available with spin transitions generated by various stimuli.

Depending on the interactions intensity among the molecules and the material characteristics, a first order LS-HS transition can take different shapes with and without hysteresis. The most common stimulus used in producing the spin transition in SCO is the variation of temperature (i.e. the intensity of thermal noise). According to the type and the intensity of the interactions, the SCO compounds present different magnetic behaviors with the temperature variation: gradual, discontinued transition (with hysteresis), two-step transition with and without hysteresis, etc. [100] (see Fig. 3.19).

In general, gradual transitions are observed in systems with weak intermolecular interactions, where each metallic center undergoes a spin transition almost independently of its neighbors. On the other hand, the presence of hysteresis may

Fig. 3.18 Electronic diagram of the HS and LS states for a Fe(II) ion in an octahedral ligand field

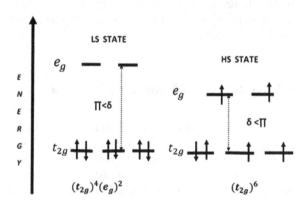

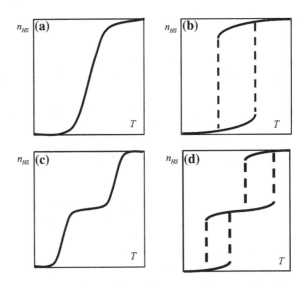

Fig. 3.19 Graphical representation of different spin transition behavior induced by variation of temperature: **a** gradual; **b** with hysteresis; **c** in two steps; and **d** in two steps with hysteresis

be associated to the presence of strong long—range interactions within the solid or with a change of crystallographic phase. The presence of the flat region in two-step type transition is generally connected to the existence of two different crystallographic sites in the system [101] or to antiferromagnetic type of the short-range interaction. These cooperative effects are reflected in the behavior of SCO materials subject to other factors, such as pressure, magnetic field or light.

The spin transition generated by the application of an external pressure can be explained by internal modification of the atomic metal–ligand distances or the crystallographic structure [102–104]. The spin transition can also be triggered by applying a magnetic field, which favors HS state of the molecule by Zeeman effect [105, 106]. A significant research interest has been devoted to the spin transitions induced by light. The light induced excited spin state trapping (LIESST) effect was first observed in 1984 on the $[Fe(ptz)_6](BF4)_2$ compound (ptz = propyltetrazole), by Decurtins et al. [107]. It was observed that the irradiation of the compound with a 514 nm light generates a transition from the fundamental state (LS) to the HS state. An explication of the phenomenon was given by using two successive intersystem sequences. The proper irradiation of LS state (term $^1A_1$) populates one of the excited levels $^1T_1$. The system is going to relax in a non radiative way to a triplet intermediary state $^3T_1$, then to the metastable HS state $^5T_2$ (or returns in the fundamental state $^1A_1$) [108].

In conclusion, hysteresis phenomena in molecular materials, such as spin-crossover compounds, are induced by various factors and are affected by thermal, acoustic, and electromagnetic noises. Consequently, hysteresis modeling with stochastic input has a significant importance in understanding the behavior of these materials and in designing practical applications of these materials.

3.5 Hysteresis in Electronic Systems

Hysteresis is often added in electronic systems in order to prevent unwanted rapid commutations, to compensate noise, or to digitize an analog signal. The most common and simple circuit to exhibit this property is Schmitt trigger (ST) which is characterized by two metastable states and a rectangular hysteretic loop for the relation between input voltage and output voltage [109]. When the positive transition of its output is generated by positive-going input the ST is represented by a non-inverted rectangular hysteretic loop embedded in a buffer, as seen in Fig. 3.20a, while when it is generated by negative-going input, the hysteretic symbol is inverted, as seen in Fig. 3.20b. There exist nowadays various implementation techniques of ST involving comparators, operational amplifiers, discrete transistors as well as some transistor–transistor logic circuits. More complex hysteretic electronic systems can be obtained by parallel connection of ST triggers with various thresholds, which resembles in the continuous approximation to the defining procedure for the Preisach model. Besides the interest of their own, the hysteretic electronic circuits play an important role in the experimental analysis of noise induced phenomena in hysteretic systems due to the possibility to control the noise characteristics of the input.

Thus, modern electronic noise generation provides the opportunity to control the noise strength and correlation as well as the shape of noise spectrum providing a wide range of known noisy environments in which the hysteretic systems can be investigated.

The relevance of hysteresis for electronics has significantly increased during the last 5 years due to the seminal results on memristor-type nanoelement presented by a group of researchers from Hewlett-Packard [110]. Thus the pinched hysteretic loop in the voltage-input plane has become one of the fingerprints of what appears to be the fourth basic circuit element, namely memory resistance or memristor. Such an element has been postulated by Chua in 1971 [111] based on some logical considerations on the relations between the four fundamental circuit variables used in the definition of the three classical circuit elements. Thus, current i, voltage v, charge q, and flux-linkage φ can be combined in six possible pairs, five of them leading to well-known relations in the circuit theory. By definition, i is related to q and φ is related to v. The characteristic law for resistor, inductor, and capacitor relates v and i, φ and i, as well as q and v, respectively. As Chua observed, the

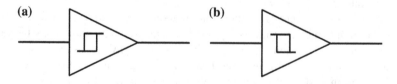

Fig. 3.20 The symbols for standard (non-inverting) Schmitt trigger (**a**) and for inverting Schmitt trigger (**b**)

relation between φ and q is missing, so it might be possible for a fourth fundamental element that is characterized by such a relation to exist (Fig. 3.21). He also
proved in [111] that the behavior of this missing element is similar to the one of a
nonlinear resistor with memory and cannot be described by combinations of the
other three fundamental elements.

Since then, several generalizations of the memristor have been proposed which
do not necessarily start from the relation between the flux linkage and charge.
They try to bring the memristor concept closer to the recent experimental findings,
especially in the nanoeletronics. Although Chua has tried to provide a unitary view
on these extensions [112, 113], the results are still debatable in the scientific
community. Nevertheless, the pinched hysteretic loop in the i-v plane became a
fingerprint of any passive electronic circuit element that adhere to the class of
memristors, independent of the physical mechanism leading to this hysteresis.

Noise is ubiquitous in electronic systems and has multiple sources, both
intrinsic and extrinsic. The voltage or current signal applied to the system bears
some external noise, while Johnson noise (due to thermal motion of the charged
particles), shot noise (resulting from the flow of current over a potential barrier) or
pink noise (whose origins are not completely understood) are often present in
electronic systems and generate a stochastic behavior of the system output.

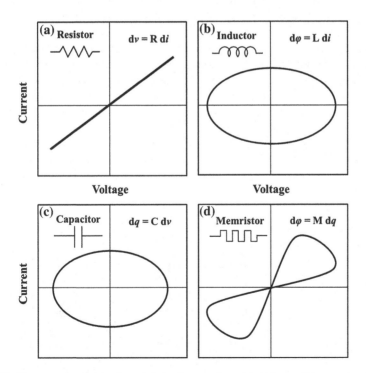

Fig. 3.21 The current–voltage characteristics for a resistor (**a**), inductor (**b**), capacitor (**c**) and
memristor (**d**) assuming a sinusoidal voltage applied to each element

3.6 Hysteresis in Wireless Communications

The hysteresis in wireless communications is mainly related to the hand-off (or handover) process [114]. In order to provide a continuous and high quality service to mobile users, a wireless communication network should be capable of changing the communication channel while a call is still in progress. This hand-off process is needed for a user moving out of the range of the current communication tower, for load balance and for interference reduction.

Let us discuss as a handoff example, the transfer made between two base stations. When a mobile user passes from the coverage area of one antenna tower to the coverage area of another antenna tower within a call's duration, the transfer should be made without interrupting the call. Intuitively one might think that the transfer should be performed when the strength of signal received by the user from the first tower becomes weaker than the strength of the signal received from the second tower. That would correspond to location **B** from the scenario depicted in Fig. 3.22. The problem with this approach to the handoff decision is related to the fluctuations of the signal received from the two antenna towers which results into many (and random) switching of the user connection with the two towers.

This ping-pong phenomenon between the towers can be avoided by using hysteresis margin for the handoff decision. Thus the transfer is performed when the average signal strength from the target tower is higher than the average signal strength from the current tower plus a given threshold (see point **C** from Fig. 3.22). Once the transfer is made smaller, the user will connect to the previous tower only when the corresponding signal becomes higher than the one received from the newly connected tower plus the given threshold. It is apparent that the ping-pong phenomenon is thus avoided. In a real wireless network the user received signals from more than two towers and the decision involves a more complex hysteretic system.

The tendency of heterogeneous wireless communications to merge into one global network has opened a new area of handoff problems in wireless

Fig. 3.22 The representation of the signal power received by a mobile user from two antenna towers when moving from one tower towards another

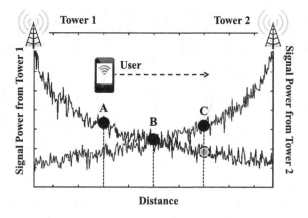

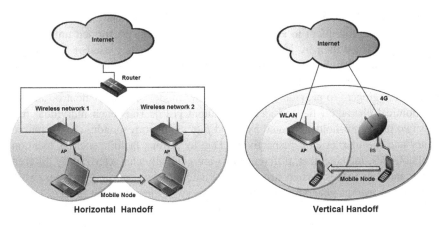

Fig. 3.23 Representation of horizontal and vertical handoff processes

communications, namely *"vertical"* handoff [115]. In the heterogeneous networks, both horizontal handoff and vertical handoff take place as illustrated in Fig. 3.23. Thus, the simultaneous existence of access technologies with different character-istics leads to a complex decision problem of determining the *best* available net-work at *best* time to perform the handoff, a problem that is still under intensive research focus [115, 116].

3.7 Economic Hysteresis

The transition from natural sciences and engineering areas to the social sciences opens the door for a more general discussion regarding the concept of hysteresis. Although the basic human intuition consider naturally that history influences the present and future behavior, the Newtonian paradigm of science has completely overturned this belief stating that the motion of an object can be predicted in terms of present quantities, while the history of the object motion is irrelevant. More-over, while this temporal action at a "distance" is not admitted, Newton has introduced the spatial action at a distance which seems also counterintuitive. The history of natural sciences proved him right but the debate about the ontological interpretation of hysteresis phenomena is still open in the social sciences. Nev-ertheless, the epistemological hysteresis is ubiquitous both in natural and social sciences and describes phenomena whose state-equations are not known or are far too complex to be used for a successful prediction.

Although ideas related to hysteresis in economic theories can be traced back until the end of nineteen century [117], the first consistent use of the term and its hypothesis is associated to Georgescu-Roegen [118, 119] in the second half of the last century. He argued that various historical factors, such as consumption experience, can influence the economic equilibrium. Although it raised significant interest in the economic community [120–122], no analytical studies to quantify hysteretic effect in economic systems have been performed until 1980's [123, 124]. Since then hysteresis modeling has been widely used in economics, especially in the studies related to unemployment, finance and international trade.

As an illustrative example, let us represent the individual agents as hysterons that can aggregate to provide a macroeconomic model of employment or another economical characteristic considered as output [125]. Thus each individual firm has two price thresholds, denoted by α and β ($\alpha > \beta$), corresponding to the market entry and the market exit, respectively. Due to various factors, such as location, technologies, management, different firms are generally characterized by different entry and exit values. When firm status is active, it has one unit of labor, while exiting the market generates zero units of labor, so the entry-exit process can be related to unemployment. Preisach hysteretic model and its generalizations (discussed in Chap. 1) are often used to describe the aggregation process of individual firms from a given market [126]. Due to the random nature of economical indicators, it is natural to consider the Preisach model driven by stochastic input [127] as a macroeconomic model of unemployment.

References

1. Wang, S. X., & Taratorin, A. M. (1999) *Magnetic information storage technology*. San Diego: Academic Press.
2. Park, K. S., Park, Y. P., & Park, N. C. (2011). Prospect of recording technologies for higher storage performance. *IEEE Transactions on Magnetics, 47*(3), 539–545.
3. Wood, R., Willlams, M., Kavcic, A., & Miles, J. (2009). The feasibility of magnetic recording at 10 terabits per square inch on conventional media. *IEEE Transactions on Magnetics, 45*, 917–923.
4. Kryder, M. H., & Gustafson, R. W. (2005). High-density perpendicular recording—advances, issues, and extensibility. *Journal of Magnetism and Magnetic Materials, 287*, 449–458.
5. Néel, L. (1949). Théorie du traînage magnétique des ferromagnétiques en grains fins avec applications aux terres cuites. *Annals of Géophysics, 5*, 99–136.
6. Weller, D., & Moser, A. (1999). Thermal effect limits in ultrahigh-density magnetic recording. *IEEE Transactions on Magnetics, 35*, 4423.
7. Rao, G. N., Yao, Y. D., & Chen, J. W. (2005). Superparamagnetic behavior of antiferromagnetic CuO nanoparticles. *IEEE Transactions on Magnetics, 41*, 3409.
8. Wernsdorfer, W. (2001). Classical and quantum magnetization reversal studies in nanometer-sized particles and clusters. *Advances in Chemical Physics, 118*, 99.
9. Coffey, W. T., Crothers, D. S. F., Dormann, J. L., et al. (1998). Thermally activated relaxation time of a single domain ferromagnetic particle subjected to a uniform field at an oblique angle to the easy axis: Comparison with experimental observations. *Physical Review Letters, 80*, 5655.
10. Stoner, E. C., & Wohlfarth, E. P. (1948). A mechanism of magnetic hysteresis in heterogeneous alloys. *Philosophical Transactions of the Royal Society, 240*, 599–642. (Reprinted in *IEEE Transactions on Magnetics, 27*(4), 3475–3518, 1991).
11. Dimian, M. (2005). *Nonlinear spin dynamics and ultra-fast precessional switchings* (1st ed.). PhD Thesis, Proquest Information and Learning.

12. Miltat, J., Albuquerque, G., & Thiaville, A. (2001). An introduction to micromagnetics in the dynamic regime. In *Spin dynamics in confined magnetic structure I* (pp. 1–31). Heidelberg: Springer.

13. Levy, L. P. (2000). *Magnetism and superconductivity*, Berlin: Springer.

14. Morrish, A. H. (1965). *The physical principles of magnetism*. New York: John Wiley and Sons.

15. Brown, W. F. (1963). *Micromagnetics*. New York: Interscience Publisher.

16. Hillebrands, B., & Ounadjela, K. (Eds.), *Spin dynamics in confined magnetic structure I*. Heidelberg: Springer.

17. Hillebrands, B., & Ounadjela, K. (2003) (Eds.), *Spin dynamics in confined magnetic structure II*. Berlin: Springer.

18. Sparks, M. (1964). *Ferromagnetic relaxation theory*. New York: McGraw-Hill Book Company.

19. Gurevich, A. G., & Melkov, G. A. (1996). *Magnetization oscillations and waves*. Boca Raton: CRC Press.

20. Landau, L., & Lifshitz, E. (1935). On the theory of the dispersion of magnetic permeability in ferromagnetic bodies. *Physik Z Sowjetunion, 8*, 135–148. (Reprinted in collected papers of L. D. Landau, Gordon and Breach, 1965).

21. Bloch, F. (1946). Nuclear inductions. *Physical Review, 70*(7), 460–474.

22. Gilbert, T. L. (2004). *Formulation, foundations and applications of the phenomenological theory of ferromagnetism*. Ph.D. Thesis, Illinois Institute of Technology, 1956; partially reprinted in IEEE Transactions on Magnetics, *40*(6), 3443–3449.

23. Kubo, R. (1966). The fluctuation-dissipation theorem. *Reports on Progress in Physics, 29*, 255–284.

24. Brown, W. F., Jr. (1963). Thermal fluctuations of a single-domain particle. *Physical Review, 130*, 1677–1686.

25. Coffey, W. T., Kalmykov, Y. P. & Waldron, J. T. (2004) *The Langevin equations with applications to stochastic problems in physics, chemistry, and electrical engineering*, Singapore: World Scientific Publishing Co Pte Ltd.

26. Rado, G. T., & Shul, H. (1963). *Magnetism III*. New York: Academic Press.

27. Aharoni, A. (1998). *Introduction to the theory of ferromagnetism*. Oxford: Clarendon Press.

28. Kronmuller, H., & Fahnle, M. (2003) *Micromagnetism and the microstructure of ferromagnetic solids*. Cambridge: Cambridge University Press.

29. Hubert, A., & Schafer, R. (1998). *Magnetic domains: The analysis of magnetic microstructures*. Berlin: Springer.

30. http://www.ctcms.nist.gov/~rdm/mumag.org.html.

31. Franchin, M., Bordignon, G., Fangohr, H., & Fischbacher, T. (2007). A Systematic Approach to Multiphysics Extensions of Finite-Element-Based Micromagnetic Simulations: Nmag. *IEEE Transactions on Magnetics, 43*(6), 2896–2898.

32. Mayergoyz, I. D., & Korman, C. E. (1991). Preisach model with stochastic input as a model for viscosity. *Journal of Applied Physics, 69*(4), 2128–2134.

33. Mayergoyz, I. D., & Korman, C. E. (1994). The Preisach model with stochastic input as a model for after effect. *Journal of Applied Physics, 75*(10), 5478–5480.

34. Korman, C. E., & Rugkwamsook, P. (1997). Identification of magnetic after effect model parameters: Comparison of experiment and simulations. *IEEE Transactions on Magnetics, 33*(5), 4176–4178.

35. Rugkwamsook, P., Korman, C. E., Bertotti, G., & Pasquale, M. (1999). Dependence of aftereffect on magnetization history. *Journal of Applied Physics, 85*(8), 4361–4363.

36. Freidlin, M. I., & Wentzell, A. D. (1993). Diffusion processes on graphs and the averaging principle. *Annals of Probability, 21*(4), 2215–2245.

37. Freidlin, M. I., & Sheu, S. J. (2000). Diffusion processes on graphs: stochastic differential equations, large deviation principle. *Probability Theory and Related Fields, 116*, 181.

38. Freidlin, M. I., Mayergoyz, I. D., & Pfeiffer, R. (Aug 2000). Noise in hysteretic systems and stochastic processes on graphs. *Physical Review E—Statistical Physics,* Plasmas, Fluids, *and Related Interdisciplinary Topics, 62* (2 Pt A), 1850–1855.
39. Mayergoyz, I. D., & Dimian, M. (2003). Analysis of spectral noise density of hysteretic systems driven by stochastic processes. *Journal of Applied Physics, 93*(10), 6826–6828.
40. Dimian, M., & Mayergoyz, I. D. (2004). Spectral noise density of the Preisach model. *IEEE Transactions on Magnetics, 40*(4), 2134–2136.
41. Dimian, M., & Mayergoyz, I. D. (Oct 2004). Spectral density analysis of nonlinear hysteretic systems. *Physical Review E, 70*(4), 046124.
42. Jiles, D. C., & Atherton, D. L. (1983). Ferromagnetic Hysteresis. *IEEE Transactions on Magnetics, 19,* 2183–2185.
43. Jiles, D. C., & Atherton, D. L. (1984). Theory of ferromagnetic hysteresis. *Journal of Applied Physics, 55,* 2115–2120.
44. Jiles, D. C., et al. (1992). Numerical determination of hysteresis parameters for the modeling of magnetic-properties using the theory of ferromagnetic hysteresis. *IEEE Transactions on Magnetics, 28,* 27–35.
45. Hauser, H. (1994). Energetic model of ferromagnetic hysteresis. *Journal of Applied Physics, 75,* 2584–2596.
46. Hauser, H. (1995). Energetic model of ferromagnetic hysteresis. 2. Magnetization calculations of (110)[001] Fesi sheets by statistic domain behavior. *Journal of Applied Physics, 77,* 2625–2633.
47. Sablik, M. J., et al. (1987). Model for the effect of tensile and compressive stress on ferromagnetic hysteresis. *Journal of Applied Physics, 61,* 3799–3801.
48. Sablik, M. J., & Jiles, D. C. (1999). Modeling the effects of torsional stress on hysteretic magnetization. *IEEE Transactions on Magnetics, 35,* 498–504.
49. Hauser, H., & Fulmek, P. (1992). The effect of mechanical-stress on the magnetization curves of Ni-single and fesi-single crystals at strong fields. *IEEE Transactions on Magnetics, 28,* 1815–1825.
50. Andrei, P., et al. (2007). Temperature, stress, and rate dependent numerical implementation of magnetization processes in phenomenological models. *Journal of Optoelectronics and Advanced Materials, 9,* 1137–1139.
51. Coleman, B. D., & Hodgdon, M. L. (1986). A constitutive relation for rate-independent hysteresis in ferromagnetically soft materials. *International Journal of Engineering Science, 24,* 897–919.
52. Coleman, B. D., & Hodgdon, M. L. (1987). On a class of constitutive relations for ferromagnetic hysteresis. *Archive for Rational Mechanics and Analysis, 99,* 375–396.
53. Hodgdon, M. L. (1991). Computation of superconductor critical current densities and magnetization curves. *Journal of Applied Physics, 69,* 4904–4906.
54. Boley, C. D., & Hodgdon, M. L. (1989). Model and simulations of hysteresis in magnetic cores. *IEEE Transactions on Magnetics, 25,* 3922–3924.
55. Hodgdon, M. L. (1988). Applications of a theory of ferromagnetic hysteresis. *IEEE Transactions on Magnetics, 24,* 218–221.
56. Hodgdon, M. L. (1988). Mathematical theory and calculations of magnetic hysteresis curves. *IEEE Transactions on Magnetics, 24,* 3120–3122.
57. Ewing, J. A. (1889). On hysteresis in the relation of strain to stress. *Report of the British Association for the Advancement of Science,* 502–504.
58. Prandtl, L. (1924). Spannungverteilung in plastischen korpern. *Proceedings of the first International Congress on Applied Mechanics* (pp. 43–54).
59. Prandtl, L. (1928). Ein gedankenmodell zur kinetischen theorie der festen korper. *Zeitschrift für Angewandte Mathematik und Mechanik, 8,* 85–106.
60. Ishlinski, A. Y (1944). Some applications of statistical methods to describing deformations of bodies (in Russian). *Izvestiya akademii nauk seriya, Tech., 9,* 580–590.
61. Wittke, H., et al. (1997). Description of stress-strain hysteresis loops with a simple approach. *International Journal of Fatigue, 19,* 141–149.

62. Wittke, H., & Rie, K. T. (1998). Phenomenological and physically based description of the deformation behaviour in cyclic plasticity. *Materials Science and Engineering a-Structural Materials Properties Microstructure and Processing, 247*, 195–203.

63. Plumtree, A., & Abdel-Raouf, H. A. (2001). Cyclic stress-strain response and substructure. *International Journal of Fatigue, 23*, 799–805.

64. Modoni, G., et al. (2011). Cyclic stress-strain response of compacted gravel. *Geotechnique, 61*, 473–485.

65. Gruning, A., et al. (2010). Cyclic stress-strain behavior and damage of tool steel AISI H11 under isothermal and thermal fatigue conditions. *Materials Science and Engineering a-Structural Materials Properties Microstructure and Processing, 527*, 1979–1985.

66. Franca, F. J. C., et al. (2010). Cyclic stress-strain response of superelastic polycrystalline Cu-12 wt%Al-0.5 wt%Be alloy. *Functional and Structural Materials, 643*, 91–97.

67. Gupta, G., et al. (2009). Cyclic stress-strain characteristics of two microalloyed steels. *Materials Science and Technology, 25*, 760–769.

68. Doring, R., et al. (2003). A plasticity model for calculating stress-strain sequences under multiaxial nonproportional cyclic loading. *Computational Materials Science, 28*, 587–596.

69. Bahn, B. Y., & Hsu, C. T. T. (Mar-Apr, 1998). Stress-strain behaviour of concrete under cyclic loading. *Aci Materials Journal, 95*, 178–193.

70. Nakai, T., et al. (1995). Stress-strain behaviour of clay under cyclic loading and its modeling. *Earthquake Geotechnical Engineering, 1 and 2*, 405–410.

71. Morrison, D. J., & Chopra, V. (1994). Cyclic stress-strain response of polycrystalline nickel. *Materials Science and Engineering a-Structural Materials Properties Microstructure and Processing, 177*, 29–42.

72. Bertotti, G., & Mayergoyz, I. (2005). *The science of hysteresis.* New York: Elsevier.

73. Gunston, T. P., et al. (2004). A comparison of two methods of simulating seat suspension dynamic performance. *Journal of Sound and Vibration, 278*, 117–134.

74. Laxalde, D., et al. (Jul–Aug, 2007). Qualitative analysis of forced response of blisks with friction ring dampers. *European Journal of Mechanics a-Solids, 26*, 676–687.

75. Laxalde, D., et al. (2006). Dynamics of a linear oscillator connected to a small strongly non-linear hysteretic absorber. *International Journal of Non-Linear Mechanics, 41*, 969–978.

76. Laxalde, D., et al. (2005) Modeling and analysis of friction rim dampers for blisks. *Proceedings of the ASME International Design Engineering Technical Conferences and Computers and Information in Engineering Conference* (vol. 1 (Pts A-C), pp. 1013–1021).

77. Kang, D. W., et al. (2010). Application of bouc-wen model to frequency-dependent nonlinear hysteretic friction damper. *Journal of Mechanical Science and Technology, 24*, 1311–1317.

78. Dominguez, A., et al. (2008). Modeling and application of MR dampers in semi-adaptive structures. *Computers & Structures, 86*, 407–415.

79. Kwok, N. M., et al. (2007). Bouc-Wen model parameter identification for a MR fluid damper using computationally efficient GA. *ISA Transactions, 46*, 167–179.

80. Ok, J. K., et al. (2008). New nonlinear bushing model for general excitations using Bouc-Wen hysteretic model. *International Journal of Automotive Technology, 9*, 183–190.

81. Rakotondrabe, M. (2011). Bouc-Wen modeling and inverse multiplicative structure to compensate hysteresis nonlinearity in piezoelectric actuators. *IEEE Transactions on Automation Science and Engineering, 8*, 428–431.

82. Sireteanu, T., et al. (2010). Identification of an extended Bouc-Wen model with application to seismic protection through hysteretic devices. *Computational Mechanics, 45*, 431–441.

83. Goda, K., et al. (2009). Probabilistic characteristics of seismic ductility demand of SDOF systems with Bouc-Wen hysteretic behavior. *Journal of Earthquake Engineering, 13*, 600–622.

84. Baber, T. T., & Noori, M. N. (1985). Random vibration of degrading, pinching systems. *Journal of Engineering Mechanics-Asce, 111*, 1010–1026.

85. Baber, T. T., & Noori, M. N. (1986). Modeling general hysteresis behavior and random vibration application. *Journal of Vibration Acoustics Stress and Reliability in Design-Transactions of the ASME, 108,* 411–420.

86. Baber, T. T., & Wen, Y. K. (1981). Random vibration of hysteretic, degrading systems. *Journal of the Engineering Mechanics Division-ASCE, 107,* 1069–1087.

87. Dobson, S., et al. (1997). Modeling and random vibration analysis of SDOF systems with asymmetric hysteresis. *International Journal of Non-Linear Mechanics, 32,* 669–680.

88. Hughes, D. (2001). The critical current of superconductors: an historical review. *Low Temperature Physics, 27*(9), 713–722.

89. Bean, C. P. (1962). Magnetization of hard superconductors. *Physical Review Letters, 8,* 250.

90. Bean, C. P. (1964). Magnetization of high-field superconductors. *Reviews of Modern Physics, 36,* 31–39.

91. Farach, H. A., Creswick, R.J., Prozorov, R., Poole, C. P. Jr., (2007). *Superconductivity* (2nd ed.). Amsterdam: Academic Press.

92. Mayergoyz, I. D. (1998). *Nonlinear Diffusion of Electromagnetic Fields: With Applications to Eddy Currents and Superconductivity,* San Diego: Academic Press.

93. Mayergoyz, I. D. (2003). *Mathematical models of hysteresis and their applications* (2nd ed.). San Diego: Academic Press.

94. Gütlich, P., Goodwin, H., & Harold, A. (Eds.). (2004). *Spin crossover in transition metal compounds I–III.* Berlin: Springer.

95. Kahn, O., & Martinez, C. J. (1988). Spin-transition polymers: from molecular materials toward memory devices. *Science, 279,* 44–48.

96. Hauser, A., Gütlich, P., & Spiering, H. (1986). High-spin to low-spin relaxation kinetics and cooperative effects in the [Fe(ptz)6](BF4)2 and [Zn(1-x)Fx(ptz)6](BF4)2 (ptz = 1-propyltetrazole) spin-crossover systems. *Inorganic Chemistry, 25*(23), 4245–4248.

97. Rotaru, A., Linares, J., Dimian, M., Nasser, J., & Gindulescu, A. (2011). Analysis of phase transitions in spin-crossover compounds by using atom-phonon coupling model. *Journal of Physics: Conference Series, 268,* 012007.

98. Gindulescu, A. (2011). *Modeling and simulation of hysteretic behavior in magnetic and molecular materials and its application to data storage.* Ph.D. thesis 2011, University of Versailles-St, Quentin and Stefan cel Mare University.

99. Cambi, L., & Szego, L. (1931). *Berichte Der Deutschen Chemischen Gesellschaft, 64,* 2591.

100. Real, J. A., Gaspar, A. B., & Muñoz, M. C. (2005). Thermal, pressure and light switchable spin-crossover materials. *Dalton Transactions, 12,* 2062–2079.

101. Tang, J. K., et al. (2009). Two-step spin-transition Iron(III) compound with a wide [high spin-low spin] plateau. *Inorganic Chemistry, 48*(5), 2128–2135.

102. Tanasa, R., et al. (2007). Piezo- and thermo- switch investigation of the spin-crossover compound [Fe(PM-BiA)$_2$(NCS)$_2$]. *Chemical Physics Letters, 443,* 435–438.

103. Long, G. J., & Hutchinson, B. B. (1987). Spin equilibrium in iron(II) poly(1-pyrazolyl)borate complexes: low-temperature and high-pressure Moessbauer spectral studies. *Inorganics Chemistry, 26*(4), 608–613.

104. Slichter, C. P., & Drickamer, H. G. (1972). Pressure-induced electronic changes in compounds of iron. *Journal of Chemical Physics, 56*(5), 2142–2160.

105. Qi, Y., Müller, E. W., Spiering, H., & Gütlich, P. (1983). The effect of a magnetic field on the high-spin α low-spin transition in [Fe(phen)$_2$(NCS)$_2$]. *Chemical Physics Letters, 101,* 503–505.

106. Bousseksou, A., et al. (2000). Dynamic triggering of a spin-transition by a pulsed magnetic field. *European Physical Journal B, 13,* 451–456.

107. Decurtins, S., Gütlich, P., Hasselbach, K. M., Hauser, A., & Spiering, H. (1985). Light-induced excited-spin-state trapping in Iron(II) spin-crossover systems. Optical spectroscopic and magnetic susceptibility study. *Inorganics Chemistry, 24,* 2174–2178.

108. Hauser, A. (1986). Reversibility of light-induced excited spin state trapping in the Fe(ptz)$_6$(BF$_4$)$_2$, and the Zn$_{1-x}$Fe$_x$(ptz)$_6$(BF$_4$)$_2$ spin-crossover systems. *Chemical Physics Letters, 124*(6), 543–548.

109. Schmitt, Otto H. (1938). A thermionic trigger. *Journal of Scientific Instruments, 15,* 24–26.
110. Snider, G. S., Stewart, D. R., Williams, R. S., & Strukov, D. B. (2008). The missing memristor found. *Nature, 453,* 80–83.
111. Chua, L. O. (1971). Memristor-the missing circuit element. *IEEE Transactions on Circuit Theory, 18,* 507–519.
112. Chua, L. C. (2011). Resistance switching memories are memristors. *Applied Physics A, 102,* 765–783.
113. Chua, L. O. (2012). The fourth element. *Proceedings of IEEE, 100,* 1920–1927.
114. Molisch, A. F. (2011). *Wireless Communications,* (2nd ed.). Chichester: John Wiley & Sons.
115. Calafate, C. T., Cano, J. C., Manzoni, P., & Barja, J. M. (2011). An overview of vertical handover techniques: Algorithms, protocols and tools. *Computer Communications, 34*(8), 985–997.
116. Sekercioglu, Y. A., Narayanan, S., & Yan, X. (2010). A survey of vertical handover decision algorithms in fourth generation heterogeneous wireless networks. *The International Journal of Computer and Telecommunications Networking, 54*(11), 1848–1863.
117. Cross, R. (1993). On the foundations of hysteresis in economic systems. *Economics and philosophy, 9,* 53–74.
118. Georgescu-Roegen, N. (1966). *Analytical economics: issues and problems.* Harvard: Harvard University Press.
119. Georgescu-Roegen, N. (1971). *The entropy law and the economic process.* Harvard: Harvard University Press.
120. Samuelson, P. A. (1965).Some notions of casuality and teleology in economics. In *Collected Scientific Papers of Paul A. Samuelson* (pp. 246–271). Cambridge: MIT Press.
121. Phelps, E. S. (1972). *Inflation policy and unemployment theory.* London: Macmillan.
122. Elster, J. (1976). A note of hysteresis in social sciences. *Synthese, 33,* 371–391.
123. Cross, R. (1988) (Ed.). *Unemployment, hysteresis and natural rate hypothesis.*Oxford: Blackwell.
124. Katzner, D. W. (1993). Some notes on the role of history and the definition of hysteresis and related concepts in economic analysis. *Journal of Post Keynesian Economics, 15*(3), 323–345.
125. Pindyck, R., Dixit, A. (1994). *Investment under uncertainty.* Princeton: Princeton University Press.
126. Cross, R., Piscitelli, L., Darby, J. (2006). Hysteresis and unemployment: a preliminary investigation. In G. Bertotti, I. Mayergoyz (Eds.), *Science of Hysteresis* (pp. 667–696). Amsterdam: Academic Press.
127. Dimian, M. (2010). *Stochastic and dynamics aspects of hysteresis* (in Romanian). Cluj-Napoca: Mediamira.

Chapter 4
Noise Driven Relaxation Phenomena in Hysteretic Systems

This chapter focuses on the description of relaxation phenomena induced by noise in hysteretic systems. First, we discuss the role of temperature in hysteretic systems and introduce the concept of thermal relaxation and its connection to rate-independent and rate-dependent hysteresis. Then, we investigate the effects of noise in various scalar and vector hysteretic systems, define scalar and vector viscosity coefficients and discuss about the data collapse phenomenon. Special consideration is given to the memory loss in the Preisach model and the degradation of the Preisach distribution as a function of time.

4.1 Temperature in Hysteretic Systems

Temperature is usually introduced as a scalar quantity that gives a measure of how hot or cold a physical system is. It relates to properties and phenomena such as the internal energy of the system, heat transfer between two bodies, and work, and stays at the basis of the laws of thermodynamics. Statistical mechanics was the first one to give a microscopic interpretation of the temperature based on the kinetic energy of individual molecules and ions that constitute the system, and advanced the fact that there is a direct connection between the total kinetic energy of individual particles and temperature. It is exactly this finding from the area of statistical mechanics that people are usually using to generalize the concept of temperature to other, often nonphysical systems. For instance, we introduce the concept of temperature in economic [1–4] and financial systems [1], in social physics [2, 3], in simulated annealing [4], and in stochastic neural networks [5, 6], in which cases we discuss about economic temperature, financial temperature, social temperature, annealing temperature, etc. In this section we introduce the concept of *temperature in hysteretic systems* (which may or may not be physical systems) and relate it to the power of the noise in the system. The temperature will be modeled by a stochastic process superimposed on the input applied to the system, as will be discussed in the next sections.

M. Dimian and P. Andrei, *Noise-Driven Phenomena in Hysteretic Systems*,
Signals and Communication Technology 218, DOI: 10.1007/978-1-4614-1374-5_4,
© Springer Science+Business Media New York 2014

First, let us understand the role of temperature in *physical systems*. It is instrumental to start this discussion by assuming that the input variable is fixed and looking at the free energy of the system as a function of some intrinsic parameters that the free energy depends on. For instance, such intrinsic parameters can be the position of a particle on a wavy surface, the position of the domain wall along the direction of wall motion in the case of bulk ferromagnetic materials, or the angle between the easy axis and the magnetization in the case of ferromagnetic nano-particles. The equilibrium states of the physical system are defined as those states which minimize the value of the free energy of the system. For simplicity, let us assume next that the energy depends only on one intrinsic parameter that we denote by θ (see Fig. 4.1) and the output variable can be computed as a function of θ, using a simple relation $y(\theta)$. The equilibrium states of the system are given by the points of local minimum represented in the figure. For instance, states denoted by "State 1" and "State 2" are both possible equilibrium points and they result in values of the output variable equal to $y(\theta_1)$ and $y(\theta_2)$, respectively.

In order for the system to change from one state to another it has to overcome certain energy barriers. The height of these energy barriers dictate the transition rates of the system from one state to another. For instance, in the case shown in Fig. 4.1, according to statistical mechanics the transition rate to pass from "State 1" to "State 2" is equal to [7–10].

$$p_{12} = \frac{1}{\tau_0} \exp\left(-\frac{w_{12}}{kT}\right), \; E\{v_t\} = 0, \, t \geq 0 \tag{4.1}$$

where w_{12} is the energy barrier that the system has to overcome to move from "State 1" to "State 2", k is Boltzmann constant, T is the absolute temperature, and τ_0 is some time constant characterizing the process. Equation (4.1) implies that any physical system has a natural tendency to shift from higher energy states to lower energy states and, one can show that, after some time, it will spend most of its time in the state with the lowest energy state, also called the ground state. Hence, we can say that the output of our physical system (which is also a hysteretic system because it can live in more than one state for the same value of the input

Fig. 4.1 The free energy of a hysteretic system as a function of some intrinsic parameter θ. Equilibrium states are the states of local minima

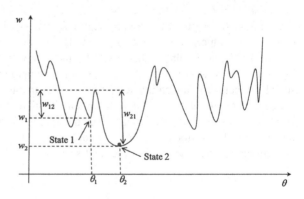

variable) is shifting in time and going towards a final state, which we call the *thermal anhysteretic state*. The higher the absolute temperature in Eq. (4.1) the faster the system is moving towards the thermal anhysteretic state. This slight change of the output of the hysteretic system from higher energy states to lower energy states is called *thermal relaxation*.

Examples of thermal relaxation processes can be found in many fields of study, such as magnetism, electronics, material science, and others [11, 12]. For instance, the magnetization of a ferromagnetic material, the polarization in ferroelectric systems, the binary state of floating gate transistors in flash memories can all change as a result of thermal relaxation.

It is now important to discuss about the connection between thermal relaxation and, rate-dependent hysteresis. Suppose we apply an input $x(t)$ to a hysteretic system and study its thermal relaxation. If the absolute temperature in (4.1) is low enough, let us say $T \approx 0$, the transition rate over any finite barrier $w_{ij} > 0$ is very small and the hysteretic system will not be able to jump over these barriers. Hence, after we apply an input x, the output of the system will change very little or not change at all because all the transition rates are negligible. When the temperature of the system is large enough so the transition rates cannot be neglected the state of the system can change even during the duration of the same experiment because of thermal relaxation, leading to a rate-dependent hysteresis phenomena.

There is a lot of work on the modeling of thermal relaxation in hysteretic systems. Most of this work was done by the magnetics community who has studied the effect of temperature and Barkhausen jumps in magnetic hysteresis. Many of the existing models for thermal relaxation are based on the Preisach model because this model can be written as a superposition of rectangular loops, for which it is relatively easy to build a relaxation model [10, 13–16]. Some other works are based on other phenomenological models such as the Coleman-Hodgdon [17], Jiles-Atherton [18], or other models of hysteresis as described in Chap. 1.

In this chapter we generalize the concept of temperature to any hysteretic system (physical and nonphysical). To increase the generality of the problem we do not look at the free energy of the system (which is usually defined only for physical systems) but simply define the temperature as a quantity proportional to the energy of fictive noise that we superimpose on the input variable. Our approach is more general than other approaches that exist in the literature, which usually can be applied to a particular hysteresis model (for instance, the approach in Ref. [17] can be applied to the Coleman-Hodgdon model, the approach in Ref. [10] can be applied to the Preisach model, etc.). Although we will discuss only about the case when the input variable is kept constant over time, the analysis in this chapter can also be applied to quasistatic input variations, i.e. when the rate of input variations is much lower than the rate of noise fluctuations. Such cases will be consider in Chap. 6 when we discuss about noise induced resonance in bistable systems.

The next three sections deal with thermal relaxation in scalar systems of hysteresis, while the last section of the chapter deals with thermal relaxation in vector systems. A few other phenomena related to thermal relaxation such as memory loss in Preisach systems and data collapse are also discussed.

4.2 Thermal Relaxation in Scalar Hysteretic Systems

In this section we consider the thermal relaxation of scalar hysteretic systems, i.e. the input and output are both scalar quantities. As it is often done in the literature [19–24], thermal fluctuations are modeled by a zero-average stochastic process v_t superimposed over the input signal x (see Fig. 4.2). Throughout this chapter we will assume that the applied input x is allowed to vary over time only for $t < 0$. For $t > 0$ the value of the applied input x is constant and equal to its value at $t = 0$. The effective value of the input is

$$x_t = x + v_t, \ E\{v_t\} = 0, \ t \geq 0, \tag{4.2}$$

where $E\{\ldots\}$ denotes the expected value of the enclosed variable. The stochastic process v_t is called *input noise* or *thermal noise* and the value x is the *holding value of the input*. Thermal relaxation occurs for $t > 0$.

If Γ is the hysteretic operator, the output variable is

$$y_t = \Gamma x_t = \Gamma(x + v_t). \tag{4.3}$$

Since v_t is a stochastic process, it is often desirable to compute the expected value of the output, which is

$$E\{y_t\} = E\{\Gamma(x + v_t)\}. \tag{4.4}$$

The type of the stochastic process v_t is given by the nature of the hysteretic system. In most cases the variance of the noise is directly proportional to the absolute temperature T. For instance, in systems of small and non-interacting magnetic particles the covariance of v_t (which, in this case, is equal to the variance of v_t) is given by [11, 25]

$$E\{v_t^2\} = \frac{2\kappa T\eta}{V}, \tag{4.5}$$

where κ is the Boltzmann constant, V is the volume of a particle, and η is the dissipation constant. Hence, by increasing (or decreasing) the temperature of a hysteretic system we mean increasing (or decreasing) the variance of the input noise.

Fig. 4.2 Modeling thermal relaxation using a noisy input

4.2.1 Monte-Carlo Algorithm for Thermal Relaxation

The expected value of the output, $E\{y_t\}$, is relatively difficult to evaluate analytically, since, in most cases, requires solving highly nonlinear and history-dependent stochastic equations. In the following we present a numerical algorithm for the computation of $E\{y_t\}$ based on Monte-Carlo simulations. This algorithm has the advantage that it is universal in the sense that it can be applied to describing thermal relaxation in the framework of any model of hysteresis such as the Preisach model, the Jiles-Atherton model, or the Energetic model and can be used for any type of noise [26].

The basic idea for the evaluation of $E\{y_t\}$ is to generate a large number of noise instances $v_{t,i}$, compute the output of the system for each such instance, and use (4.4) to evaluate the expected value of the output. This algorithm is represented in Fig. 4.3. The most computationally expensive part of the algorithm is the numerical evaluation of the output variable for each noise sample. From our experience the number of computations required to evaluate $E\{y_t\}$ varies between 10 and 1000 depending on the magnitude of the noise, on the type of the hysteresis model, and, often, on the model parameters. Usually, the computation time is a few seconds on a normal personal computer operating at 3 GHz for all the models presented in Chap 1.

In the next three subsections we present analytical and numerical results related to thermal relaxation in scalar models of hysteresis.

4.2.2 Thermal Relaxation of One Hysteron

In this section we consider the thermal relaxation of one hysteron with critical fields α and β ($\beta \leq \alpha$), and compute the expected value of the output as a function of time. This problem has been solved analytically by Mayergoyz [27] for the cases when thermal noise v_t is described by a discrete time i.i.d. process or by an Ornstein–Uhlenbeck random process. When the thermal noise is modeled by a

Fig. 4.3 Monte-Carlo algorithm for the evaluation of $E\{y_t\}$

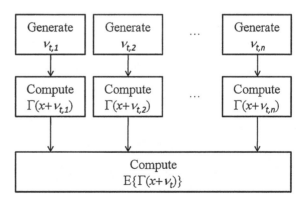

discrete time i.i.d. process the analytical derivation of the expected value of the output variable is relatively straightforward. A derivation similar to the one in [27] is presented in the next paragraphs. When the thermal noise is an Ornstein–Uhlenbeck random process the computation of $E\{\hat{\gamma}_{\alpha\beta}x_t\}$ is more complicated as it involves the solution of the "exit problem" [28–30]. In this case $E\{\hat{\gamma}_{\alpha\beta}x_t\}$ can be expressed semi-analytically in terms of the Laplace transforms and parabolic cylinder functions. A derivation of the statistical properties of the output in the case of the Ornstein–Uhlenbeck noise can be found in [27, 30].

If the noise is an i.i.d. random process the expected value of the output is

$$
\begin{aligned}
E\{\hat{\gamma}_{\alpha\beta}x_t\} &= P\{\hat{\gamma}_{\alpha\beta}x_t = 1\} - P\{\hat{\gamma}_{\alpha\beta}x_t = -1\} = 2P\{\hat{\gamma}_{\alpha\beta}x_t = 1\} - 1 \\
&= 2\left(P_{\alpha\beta}^{++}P\{\hat{\gamma}_{\alpha\beta}x_{t-1} = 1\} + P_{\alpha\beta}^{-+}P\{\hat{\gamma}_{\alpha\beta}x_{t-1} = -1\}\right) - 1 \\
&= 2\left(P_{\alpha\beta}^{++} - P_{\alpha\beta}^{-+}\right)P\{\hat{\gamma}_{\alpha\beta}x_{t-1} = 1\} + 2P_{\alpha\beta}^{-+} - 1 \\
&= \left(P_{\alpha\beta}^{++} - P_{\alpha\beta}^{-+}\right)\left(E\{\hat{\gamma}_{\alpha\beta}x_{t-1}\} + 1\right) + 2P_{\alpha\beta}^{-+} - 1,
\end{aligned}
\tag{4.6}
$$

where we have considered that time t is discrete ($t = 0, 1, 2,...$) and used the total probability theorem. Quantities $P_{\alpha\beta}^{-+}$ and $P_{\alpha\beta}^{++}$ are the probabilities for the hysteron to switch from the -1 and $+1$ and to remain in the $+1$ state, from time t-1 to t. These probabilities can be calculated as:

$$
P_{\alpha\beta}^{++} = \int_{\beta}^{\infty} \rho(x)dx,
\tag{4.7}
$$

$$
P_{\alpha\beta}^{-+} = \int_{\alpha}^{\infty} \rho(x)dx,
\tag{4.8}
$$

where $\rho(x)$ is the probability density function of the i.i.d. input noise. Equation (4.6) is a recurrence equation, which can be solved for $E\{\hat{\gamma}_{\alpha\beta}x_t\}$:

$$
E\{\hat{\gamma}_{\alpha\beta}x_t\} = \left(P_{\alpha\beta}^{++} - P_{\alpha\beta}^{-+}\right)^t\left(y_{\alpha\beta}^0 - y_{\alpha\beta}^\infty\right) + y_{\alpha\beta}^\infty,
\tag{4.9}
$$

where $y_{\alpha\beta}^0 = E\{\hat{\gamma}_{\alpha\beta}x_0\} = \hat{\gamma}_{\alpha\beta}x_0$ is the initial value of the output variable, which can be either $+1$ or -1 and

$$
y_{\alpha\beta}^\infty = \frac{P_{\alpha\beta}^{-+} + P_{\alpha\beta}^{++} - 1}{P_{\alpha\beta}^{-+} - P_{\alpha\beta}^{++} + 1}
\tag{4.10}
$$

is the limiting value of the output when $t \to \infty$.

In the special case when $\alpha = \beta$

$$
E\{\hat{\gamma}_{\alpha\alpha}x_t\} = y_{\alpha\alpha}^\infty = 2P_{\alpha\alpha}^{-+} - 1.
\tag{4.11}
$$

Figure 4.4 presents the expected value of the output of one hysteron when the noise is Gaussian and Ornstein–Uhlenbeck with different magnitudes. The initial

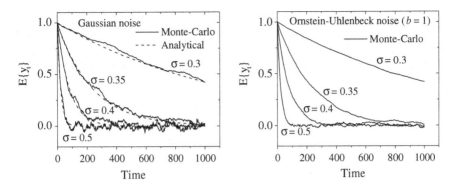

Fig. 4.4 Expected value of the output of one hysteron in the presence of Gaussian (*left*) and Ornstein–Uhlenbeck (*right*) noises. The dashed lines show the results computed using (4.9)

state is $y_{\alpha\beta}^0 = +1$ and the critical fields of the hysteron are $\alpha = 1$ and $\beta = -1$. The analytical results obtained by using Eq. (4.9) in the case of Gaussian noise are represented with dashed line. We note the good agreement between the analytical and the Monte-Carlo results. The expected value of the output decreases monotonically to 0 as a function of time in every case. At $t = \infty$ the hysteron loses completely the memory of its initial state.

4.2.3 Thermal Relaxation in the Preisach Model

In the case of the Preisach model the expected value of the output can be computed using superposition [27]. Indeed, taking the expected value in (1.16) we obtain that

$$E\{y_t\} = \iint_{\alpha \geq \beta} P(\alpha, \beta)E\{\gamma_{\alpha\beta}x_t\} \, d\alpha \, d\beta + \int_{-\infty}^{\infty} R(\alpha)E\{\hat{\gamma}_{\alpha\alpha}x_t\} \, d\alpha \qquad (4.12)$$

where $P(a, \beta)$ and $R(\alpha)$ are the irreversible and reversible components of the Preisach distribution function. Equation (4.12) shows that the expected value of the output can be computed using the Preisach model, in which the Preisach distribution is replaced with the "effective" Preisach distribution

$$E\{y_t\} = \iint_{\alpha \geq \beta} P_x^{eff}(\alpha, \beta) \, d\alpha \, d\beta + \int_{-\infty}^{\infty} R_x^{eff}(\alpha)d\alpha \qquad (4.13)$$

where

$$P_x^{eff}(\alpha, \beta) = P(\alpha, \beta)E\{\gamma_{\alpha\beta}x_t\} \qquad (4.14)$$

$$R_x^{eff}(\alpha) = R(\alpha)E\{\hat{\gamma}_{\alpha\alpha}x_t\} \qquad (4.15)$$

The irreversible and reversible components of the effective Preisach distribution $P_x^{eff}(\alpha, \beta)$ and $R_x^{eff}(\alpha)$ depend on the holding value of the input x but are not dependent on the history of the hysteretic system.

If the input noise is i.i.d. the expected value of the output at (discrete) time t is:

$$E\{y_t\} = \iint_{\alpha \geq \beta} P(\alpha, \beta)\left[\left(P_{\alpha\beta}^{++} - P_{\alpha\beta}^{-+}\right)^t\left(y_{\alpha\beta}^0 - y_{\alpha\beta}^\infty\right) + y_{\alpha\beta}^\infty\right] d\alpha\, d\beta$$
$$+ \int_{-\infty}^{\infty} R(\alpha)\left(2P_{\alpha\alpha}^{-+} - 1\right)d\alpha, \tag{4.16}$$

and the irreversible and reversible components of effective Preisach distribution are

$$P_x^{eff}(\alpha, \beta) = P(\alpha, \beta)\left[\left(P_{\alpha\beta}^{++} - P_{\alpha\beta}^{-+}\right)^t\left(y_{\alpha\beta}^0 - y_{\alpha\beta}^\infty\right) + y_{\alpha\beta}^\infty\right] \tag{4.17}$$

$$R_x^{eff}(\alpha) = R(\alpha)\left(2P_{\alpha\alpha}^{-+} - 1\right) \tag{4.18}$$

where $P_{\alpha\beta}^{-+}$ and $P_{\alpha\beta}^{++}$ are given by (4.7) and (4.8). $y_{\alpha\beta}^0$ is equal to $+1$ or -1, depending on the initial state ("up" or "down") of the elementary hysteresis loop with switching values α and β.

Figure 4.5 presents sample relaxation results when the input noise is Gaussian and Ornstein–Uhlenbeck. The irreversible component of the Preisach distribution is assumed to be normal along the interaction axis and lognormal along the coercivity axis [see Eq. (1.24)] with $\eta_0 = 1$, $\sigma_\eta = 0.3$, $\sigma_\xi = 0.3$; the reversible component is normal with $\sigma_R = 0.3$, the saturation $y_{sat} = 1$ and $S = 0.9$. The initial state for the simulations represented in Fig 4.5 is obtained by first saturating the systems and, then, bringing the input to 0 (to the remanent state). Figure 4.6 presents similar simulation results obtained by starting from a different initial state. This initial state is reached by starting from the zero-field anhysteretic state and applying an input equal to the coercive field (x_C).

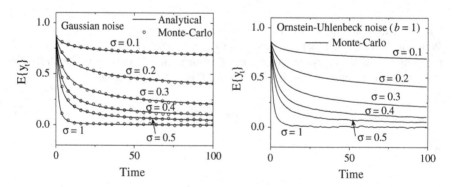

Fig. 4.5 Thermal relaxation in the case of the Preisach model when the input noise is Gaussian (*left*) and Ornstein–Uhlenbeck (*right*) distributed. The analytical results are computed using (4.16). The initial state of the system is the remanent state

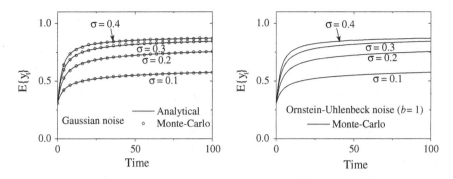

Fig. 4.6 Thermal relaxation in the case of the Preisach model when the input noise is Gaussian (*left*) and Ornstein–Uhlenbeck (*right*) distributed. The analytical results are computed using (4.16). The initial state of the system is obtained by starting from the zero-anhysteretic state and applying an input equal to the value of the coercive field ($x = x_C$)

4.2.4 Thermal Relaxation in the Framework of Other Models of Hysteresis

It is difficult to derive analytical expressions for the expected value of the output in the case of most models of hysteresis. This task was somewhat simplified in the case of the Preisach model because the model could be expressed as a linear superposition of elementary rectangular hysteresis operators, however, it is practically impossible to derive analytical expressions for the expected value of the output in the case of the other models of hysteresis presented in Chap. 1. In this subsection we present a few simulation results for thermal relaxation in the case of the Jiles-Atherton model, Coleman-Hodgdon model, energetic model, and Bouc-Wen models. It is shown that all these models of hysteresis predict qualitatively similar output relaxation behaviors.

In order to compare the results obtained using different models we have selected the parameters of the Jiles-Atherton, energetic, and Coleman-Hodgdon models in such a way to describe approximately a major hysteresis loop with saturation $y_{sat} \approx 1$, coercivity $x_c \approx 1$, and remanence $y_r \approx 0.4$. In the case of the Bouc-Wen model we have imposed $y_{sat} \approx 1$. The model parameters are given below:

1. Jiles-Atherton model: $a = 1.3$, $\alpha = 2$, $c = 0.1$, and $k = 1.2$.
2. Energetic model: $c_r = 0.1$, $g = 8$, $h = 0.1$, $k = 0.1$, $N_e = 2$, and $q = 10$.
3. Coleman-Hodgdon model: $A_1 = 1$, $A_2 = 1.6$, $A_3 = -1.3$, $A_4 = 0.3$, $\alpha = 1$, $y_{bp} = y_{cl} = 1$.
4. Bouc-Wen model: $A = 6$, $\alpha = 0$, $\beta = 0.5$, $D = 1$, $\gamma = 0.1$, $k = 1$, and $n = 1.1$.

The initial state for the first three models was obtained by applying an input of $x = 20$ to saturate the system and, then, setting $x = 0$ to arrive at the remanent point. In the case of the Bouc-Wen model (since it describes clockwise hysteresis) we have applied an input of $x = -20$ and, then, set $x = 0$. The results of the simulations are presented in Fig. 4.7. The thermal noise was always chosen Gaussian with the magnitude indicated in each figure. It is remarkable that all models considered in this study predict qualitatively similar results. The small differences come from the fact that the models are intrinsically built to describe different output dynamics for the same input process.

Figure 4.8 presents results for the thermal relaxation in the case of the Coleman-Hodgdon model for different input histories with the same holding value of the input. Note that the value of the output converges towards the same final value, which shows that the effect of the input noise is to erase the past history of the system. This phenomenon is universal in the sense that it can be observed for all the models of hysteresis used in this book. Even vector models of hysteresis, as we will see in Sect. 4.5.1, lose their past history in the presence of thermal noise. In the next section we analyze the memory loss in hysteretic systems in more details by looking at the Preisach model.

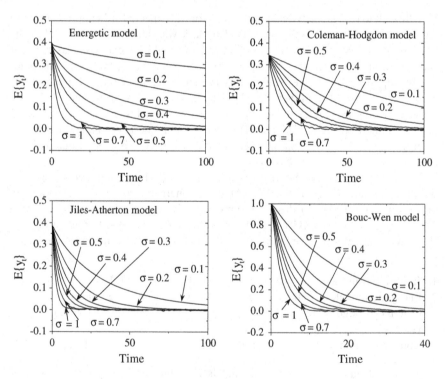

Fig. 4.7 Thermal relaxation in the case of the energetic, Jiles-Atherton, Coleman-Hodgdon, and Bouc-Wen models of hysteresis. Gaussian input noise is assumed in all cases

Fig. 4.8 Thermal relaxation
in the case of the Coleman-
Hodgdon model for the same
holding value of the input but
for different input histories

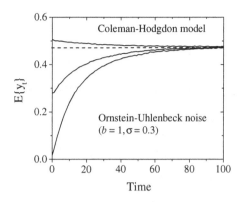

4.3 Memory Loss in Preisach Systems

An interesting phenomenon observed during the thermal relaxation of hysteretic
systems is the gradual loss of memory of the initial state with time. As shown in
Chap. 1 the memory of the Preisach model is stored in the shape of the line of
separation between positive and negative hysterons. At time $t = 0$ the line of
separation (the "stairline") is well-defined. The hysterons above this line are all in
the -1 state, while the hysterons below are in the $+1$ state. After $t = 0$ the line of
separation can change its position and the state of the hysterons above and below
this line becomes nondeterministic.

In order to analyze the loss of memory in the Preisach model we use the Monte-
Carlo method to simulate a large number of relaxations and take the average of the
state of each hysteron at the end of the simulation [31]. In this way we compute
the expected value of each hysteron in the Preisach plane for any initial state.
Theresults of these simulations are presented in Figs. 4.9, 4.10, and 4.11 for different initial states and for different types of noise. To graphically represent the loss
of memory we compute the absolute value of each hysteron in the Preisach plane
$\left| E\{\hat{\gamma}_{\alpha\beta} x_t\} \right|$ as well as the effective Preisach distribution $P_x^{eff}(\alpha, \beta)$ given by
Eq. (4.14). Quantity $\left| E\{\hat{\gamma}_{\alpha\beta} x_t\} \right|$ is close to 1 when the hysteron is either in the $+1$ or
-1 state with high probability; the closer the value of $\left| E\{\hat{\gamma}_{\alpha\beta} x_t\} \right|$ is to 0 the more
likely the hysteron has lost the memory of its initial state. At $t = 0$ all hysterons are
either in the -1 or $+1$ states and the "stairline" is well-defined.

Figure 4.9 shows the separation line and the irreversible component of the
effective Preisach distribution in the case of a Gaussian noise with increasing
standard deviation when the initial state is the zero-field anhysteretic state [obtained
by applying an alternative input with decreasing magnitude (see Chap. 1)]. During
thermal relaxation the "stairline" is becoming less defined, which is indicated
by the broadening of the separation line on the figures on the left in Fig. 4.9.
The positive and negative region of the Preisach distribution lose their deterministic
character as shown on the figures on the right.

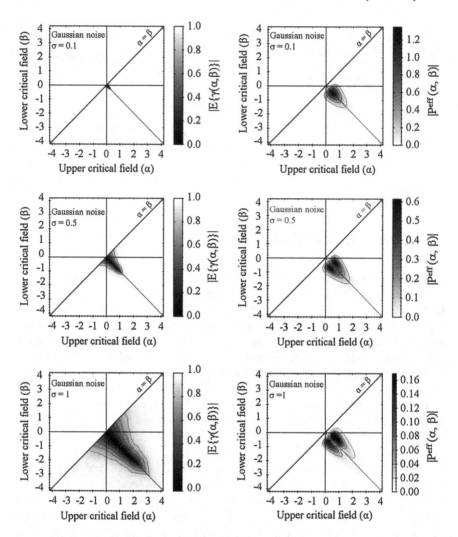

Fig. 4.9 (*Left figures*) Degradation of the separation line between the +1 and −1 hysterons during memory loss for different standard deviations of the Gaussian noise. (*Right figures*) The degradation of the positive and negative regions of the irreversible component of the effective Preisach distribution during memory loss

Figure 4.10 shows the broadening of the separation line and the irreversible component of the effective Preisach distribution when the input noise is Cauchy and Ornstein–Uhlenbeck. Note that in all cases the separation line is losing its deterministic character and the state of the hysterons in the Preisach plane is becoming uncertain with time.

Finally, Fig. 4.11 shows the separation line and the Preisach distribution when the initial state is obtained as follows: first we apply a strong input to saturate the

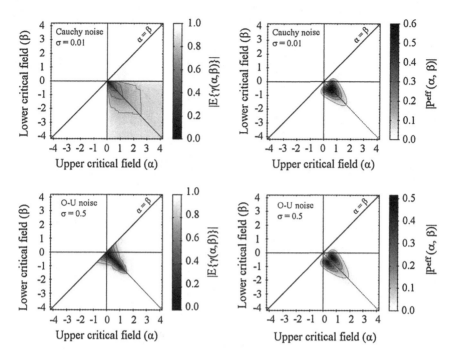

Fig. 4.10 (*Left figures*) Degradation of the separation line between the +1 and −1 hysterons during memory loss for in the case of Cauchy and Ornstein–Uhlenbeck noises. (*Right figures*) The degradation of the positive and negative regions of the irreversible component of the effective Preisach distribution during memory loss

system, then we apply an input equal to minus the coercive field ($x = -x_C$), and finally set the input to $x = x_C$. In these simulations the noise is Ornstein–Uhlenbeck with $b = 1$ and the standard deviation is indicated on each figure. Note again the broadening of the separation line between the hysterons in the +1 state and those in the −1 state. The higher the temperature the faster the separation line loses its initial shape and becomes parallel to line $\beta = -\alpha$, which is separation line in the anhysteretic state. For this reason, it has often be suggested to obtain the zero-field anhysteretic state by setting the input to 0 and keeping the system at a high temperature. Although this might produce similar results, the two techniques are fundamentally different. The zero-field anhysteretic state is a well defined state in which all hysterons are either in the −1 or +1 state. By keeping the system at high temperature for a long time, the state of the system loses its deterministic character and the hysterons are found randomly in the +1 or −1 state in such a way that the total output averages to 0. As soon as the thermal noise is removed the hysterons with $\alpha < x$ will be in the +1 state the ones with $\beta > x$ will be in the −1 state. The hysterons with $\beta < x < \alpha$ will keep their nondeterministic state.

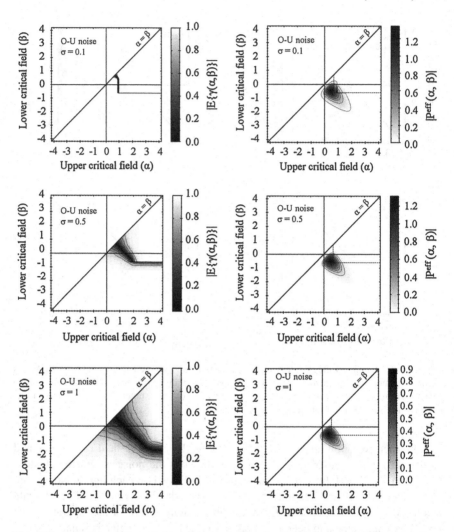

Fig. 4.11 (*Left figures*) Degradation of the separation line between the +1 and −1 hysterons during memory loss for in the case of Ornstein–Uhlenbeck noise. (*Right figures*) The degradation of the positive and negative regions of the irreversible component of the effective Preisach distribution during memory loss

4.4 Viscosity Coefficients and Data Collapse Phenomena

The expected value of the output $E\{y_t\}$ might often be approximated to change approximately logarithmically in time. In such cases, it is relevant to introduce the viscosity coefficient, defined as

$$S = \frac{dE\{y_t\}}{d \ln t} \qquad (4.19)$$

This viscosity coefficient depends mainly on the past history of the system but it can also depend on time. If the expected value of output $E\{y_t\}$ increases in time than $S > 0$, if $E\{y_t\}$ decreases than $S < 0$. To illustrate how the viscosity coefficient is computed in applications we present the time-dependence of the expected value of the output as a function of the holding value of the input variable in Fig. 4.12a and the corresponding values of the viscosity coefficient in Fig. 4.12b in the case of the Preisach model. The initial state in these simulations is obtained by applying a strong input to bring the system to positive saturation and, then, decrease the input to the holding value x (shown on the x-axis in Fig. 4.12b). The analytical results obtained by using Eqs. (4.16) and (4.19) are represented by continuous lines, while the numerical results obtained after Monte-Carlo averaging are represented by symbols. The noise is assumed Gaussian distributed in all cases. Note that the viscosity coefficient has one maximum around the value of the coercive field.

Figure 4.13 presents results obtained with the Preisach, Jiles-Atherton, Coleman-Hodgdon, and energetic models of hysteresis, when the initial state is the zero-field anhysteretic state. Notice the qualitative agreement between all the models of hysteresis used in these simulations. The slight quantitative disagreement comes from the different behavior of the models to the same input signal.

An interesting property of the viscosity coefficient is the *data collapse property* observed by Mayergoyz [33, 34] in the framework of the Preisach model of hysteresis. The data collapse property refers to the universality of following factorization of the viscosity coefficient:

$$S(x, T) = S_0 \left[\frac{x}{f(T)} \right] g(T) \qquad (4.20)$$

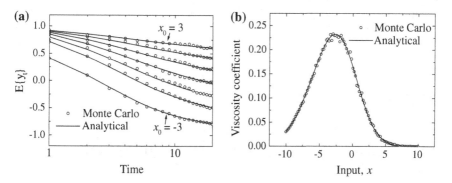

Fig. 4.12 **a** Decrease of the expected value of the output in the case of the Preisach model for different holding values x_0 of the input. The holding values of the input increase from -3 to 3 in steps of 1. **b** Viscosity coefficient as a function of the holding value of the input. (© 2008 IEEE, [32])

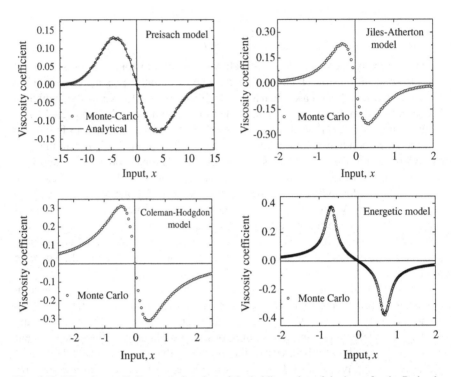

Fig. 4.13 Viscosity coefficients as a function of the holding value of the input for the Preisach, Jiles-Atherton, Hodgdon, and energetic models. The initial state is the zero-field anhysteretic state. (© 2008 IEEE, [32])

where x is the holding value of the input at which the thermal relaxation is observed, and f, g, and S_0 are some functions. Practically, if we represent the viscosity coefficient as a function of x for different noise magnitudes, all curves collapse into one universal curve S_0 after appropriate scaling. Such an example is presented in Fig. 4.14, in which we show the viscosity coefficient as a function of the applied field for different noise magnitudes. These curves have been computed by using the Preisach model. The data collapse phenomenon has been verified by both theoretical computations and experimental measurements [27] using scalar models.

4.5 Thermal Relaxation in Vector Hysteretic Systems

In this section the analysis of thermal relaxation is generalized to multidimensional hysteretic systems. First, we analyze the displacement of the output vector, after which we define and analyze the viscosity coefficients.

Fig. 4.14 Data collapse phenomenon simulated using the Preisach model. (© 2008 IEEE, [32])

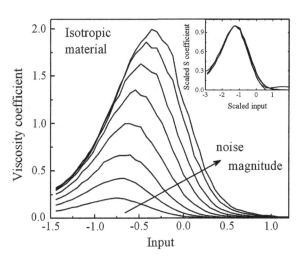

4.5.1 Output Displacement in Vector Systems

The analysis presented in Sect. 4.2 can be extended to multidimensional (vector) hysteretic systems. In these systems the deterministic input and the input noise are vector quantities

$$x_t = x + v_t, \; E\{v_t\} = 0 \tag{4.21}$$

If Γ is the vector hysteretic operator, the output variable can be computed as

$$y_t = \Gamma x_t = \Gamma(x + v_t) \tag{4.22}$$

and the expected value of the output is

$$E\{y_t\} = E\{\Gamma(x + v_t)\} \tag{4.23}$$

The Monte-Carlo algorithm presented in Fig. 4.3 for the evaluation of $E\{y_t\}$ can be generalized to multidimensional hysteresis systems. To illustrate the behavior of the expected value of the output, next, we present sample simulation results for isotropic two-dimensional systems. The simulations are performed by using the energetic, Jiles-Atherton, and Preisach models in two dimensions. In each case we consider three initial states of the hysteretic system, labeled by "A", "B", and "C":

"State A" First we bring the system to the zero-field anhysteretic state by applying a strong rotating and decreasing input like in Fig. 4.15a, after which we apply a constant input $x = (x_x, x_y)$ and wait for thermal relaxation (see Fig. 4.15b).

"State B" We bring again the system to the zero-field anhysteretic state by applying a strong rotating and decreasing input. Then, we apply an input $x = (0, x_y)$, change it to $x = (x_x, x_y)$, and wait for thermal relaxation (see Fig. 4.15c).

"State C" First we saturated the system by applying a strong input in the x-direction, after which we bring the field to $x = (0, x_y)$, change it to $x = (x_x, x_y)$, and wait for thermal relaxation (see Fig. 4.15d).

The results of the simulations are presented in Fig. 4.16. In all cases we observe that the trajectory of the output vector is linear and converges towards the same final point. The coordinates of the final point do not depend on the initial state of the system. This important property is similar to the property of scalar systems of hysteretic in which the output is converging towards the same final state (see Fig. 4.8).

In the case of anisotropic systems the trajectory of the output vector during thermal relaxation is not always a straight line as shown in the simulations presented in Fig. 4.17. The simulations were performed by using the vector Preisach model for an anisotropic system with coercive fields of 0.6 along the easy axis (the x- axis) and 3 along the hard axis (the y-axis). Note that, although the output vector was initially oriented in the positive direction, due to thermal agitation, it has eventually switched its orientation in the opposite direction. Such curved trajectories can be observed particularly when the output vector goes from one easy direction to another easy direction of the hysteretic system.

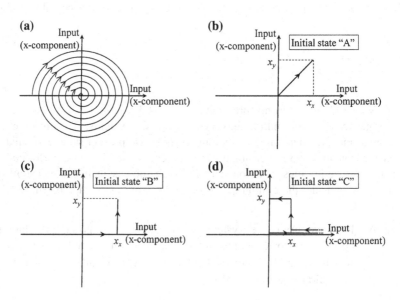

Fig. 4.15 a Obtaining the zero-field anhysteretic state in two dimensions. **b–d** Diagrams showing how the initial states labeled "A", "B", and "C" are obtained

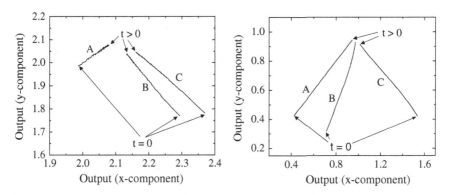

Fig. 4.16 Examples of thermal relaxations in two-dimensional hysteresis systems. (© 2008 AIP, [35])

One can also see that, in anisotropic systems, the output vector can cross the direction of the input vector. This is a pure anisotropic phenomenon, which is not met in the case of isotropic systems. In the simulations presented in Fig. 4.16 the output vector was moving to the direction of the applied field, but never crossed it.

4.5.2 Viscosity Coefficients in Vector Systems

The definition of the viscosity coefficient (4.19) can be applied only to hysteretic systems in which the direction of the output vector is constant. This usually happens in (a) isotropic systems and (b) when the direction of the input is fixed. If any of the two conditions are not satisfied the output vector might change its direction in time and (4.19) cannot be used. This discussion prompts us to look for a more general definition of the viscosity coefficient that can be reduced to the scalar one in the case of one-dimensional systems and, if possible, preserve the data collapse property introduced in Sect. 4.4.

The generalization of the scalar viscosity coefficient to multidimensional systems is challenging particularly because the trajectory of the output can deviate substantially from a straight line in vector systems. For instance, as shown in the previous section, the output variable can change direction and the path described by the output vector can be curved (see Fig. 4.17b). One way to generalize the viscosity coefficient to vector systems is to consider that (4.19) can be written for each component of the output, i.e.:

$$y_i(t) \sim y_i(0) + S_i(\mathbf{x}) \log t, \tag{4.24}$$

where i the index of component ($i = x, y, z$ in the case of three-dimensional systems) and $S_i(\mathbf{x}, T)$ are the viscosity coefficients corresponding to each axis. However, if we consider that (4.24) is valid in each dimension one can show that

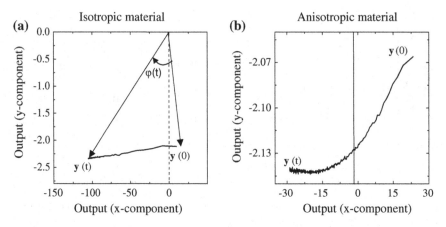

Fig. 4.17 Thermal relaxations in the cases of (**a**) isotropic and (**b**) anisotropic hysteretic systems. (© 2008 AIP, [35])

the trajectory of the output should always be a straight line, which is not true for many anisotropic materials (see Fig. 4.17b). Indeed, by considering two different axes, i and j, the slope of the projection of the output on the plane (i, j) is

$$\frac{dy_i}{dy_j} = \frac{S_i(\boldsymbol{x})}{S_j(\boldsymbol{x})}, \tag{4.25}$$

which is time-independent. The only way to make the trajectory of the output not to follow a straight line would be to consider that the viscosity coefficients along each direction are time-dependent.

A better approach to generalize the scalar viscosity coefficient to vector systems is to look at the total length of the trajectory described by the output vector [32]. For instance, in the case of three-dimensional system, where the output is $\boldsymbol{y} = (y_x, y_y, y_z)$, the total length of the trajectory described by the output vector from time t_1 till time t_2 is:

$$\Delta y(t_1, t_2) = \int_{t_1}^{t_2} dy = \int_{t_1}^{t_2} \sqrt{dy_x^2 + dy_y^2 + dy_z^2}$$

$$= \int_{t_1}^{t_2} \sqrt{\left(\frac{dy_x}{dt}\right)^2 + \left(\frac{dy_y}{dt}\right)^2 + \left(\frac{dy_z}{dt}\right)^2} \, dt \tag{4.26}$$

By introducing (4.24) into (4.26) we obtain that

$$\Delta y(t_1, t_2) = S_{3D}(\boldsymbol{x}) \log \frac{t_2}{t_1}, \quad t_2 > t_1, \tag{4.27}$$

where

$$S_{3D}(x) = \sqrt{S_x^2(x) + S_y^2(x) + S_z^2(x)} \tag{4.28}$$

is the "effective" viscosity coefficient. Equation (4.27) is the vector generalization of the scalar definition (4.19).

In order to test if the total length of the trajectory described by the output vector can be approximated as depending logarithmically in time we plot $\Delta y(t_1, t_2)$ as a function of time, t_2 on a linear scale in Fig. 4.18 and on a logarithmic scale in the inset of this figure. It is apparent that $\Delta y(t_1, t_2)$ depends more or less logarithmically on time, which supports the definition given in (4.27).

Next, we present the results of simulations for the viscosity coefficient and data collapse phenomena in vector hysteretic systems. We have performed the following two types of simulation experiments:

Simulation 1. First we saturate the system in the positive x-direction, then we apply an input x_i along the x-axis, followed by a holding field x_i in the y-direction, (x_i, x_i); then we wait for thermal relaxation (see Fig. 4.19a).

Simulation 2. We come again from positive saturation to an input x_i along the x-axis, after which we apply a field of the same magnitude x_i along the y-direction, and finally remove the field along the x-direction, so that the final holding value of the field is $(0, x_i)$; then, we wait for thermal relaxation (see Fig. 4.19b).

These simulations were performed for values of x_i ranging from positive to negative saturation. Each time the viscosity coefficient was obtained using (4.28), where x is the holding value of the input. The first type of experiment corresponds

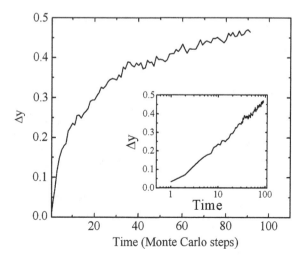

Fig. 4.18 Absolute length of the trajectory described by the output vector during thermal relaxation on a linear scale and on a log scale (*the inset*). (© 2008 IEEE, [32])

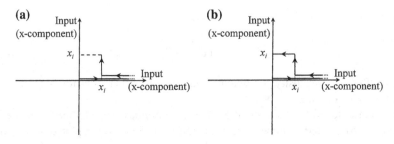

Fig. 4.19 Procedures used to obtain the initial state of the 2-dimensional hysteretic system

to a first-order rotational process, while the second type corresponds to a second-order process [36].

The viscosity coefficients obtained by using the first type of simulations ("Simulation 1") are represented in Fig. 4.20 for different noise magnitudes. The inset of this figure shows the scaled curves of the viscosity coefficients. It is remarkable that even for this relatively simple type of experiment we obtain two peaks for the viscosity curves. This is a purely vectorial effect that can be attributed to rotational changes of the output vector during relaxation. It is also interesting to observe that the two peaks correspond to values of the input that are close to critical fields of the system, which in our simulations are approximately ±0.6. All the viscosity curves seem to scale into one universal curve just like in the case of scalar relaxation phenomena.

In Fig. 4.21 we have represented the viscosity coefficients obtained using the second type simulations ("Simulation 2") at the same magnitudes of the noise. The viscosity coefficient curves show two or even three maxima, depending on the values of the noise magnitudes. At low noise magnitudes, the viscosity coefficient curves show only two maxima because, during the duration of the simulation

Fig. 4.20 Viscosity coefficient as a function of the input for different noise magnitudes: (from *bottom* to *top*) 0.25, 0.5, 0.8, 1.2, 1.6, 2, 2.5, and 3. The initial state is obtained as shown in Fig. 4.19a. The *inset* shows the data collapse property. (© 2008 IEEE, [32])

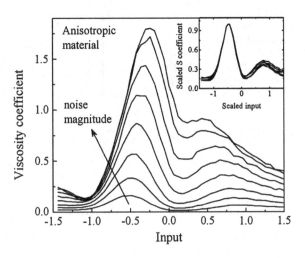

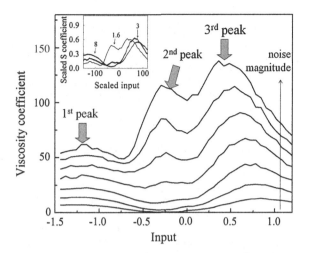

Fig. 4.21 Viscosity coefficient as a function of the value of the input for different noise magnitudes: (from *bottom* to *top*) 0.25, 0.5, 0.8, 1.2, 1.6, 2, 2.5, and 3. The initial state is obtained as shown in Fig. 4.19b. (© 2008 IEEE, [32])

experiment the thermal agitation is not strong enough to clear the system history completely. At high noise magnitudes the viscosity coefficient curves show three maxima, which correspond to the number of reversal points and the holding field in Fig. 4.19b. It is apparent from these simulations that the more complex the history of the system is the more maxima can appear on the viscosity curves. In addition, the data collapse phenomena are more difficult to observe for higher rotational reversal-curves.

References

1. Kozuki, N., & Fuchikami, N. (2003). Dynamical model of financial markets: fluctuating 'temperature' causes intermittent behavior of price changes. *Physica a-Statistical Mechanics and Its Applications, 329*, 222–230.
2. Krause, S. M., & Bornholdt, S. (Nov 2012). Opinion formation model for markets with a social temperature and fear. *Physical review. E, Statistical, nonlinear, and soft matter physics, 86*, 056106.
3. IJzerman, H., & Semin, G. R. (2010). Temperature perceptions as a ground for social proximity. *Journal of Experimental Social Psychology, 46*, 867–873.
4. Laarhoven, P. J. M., & Aarts, E. H. (1987). *Simulated annealing: Theory and applications*. New York: Springer.
5. Kirkpatrick, S. et al. (1983, May 13). Optimization by simulated annealing. *Science, 220*, 671–680.
6. Černý, V. (1985, Jan 01). Thermodynamical approach to the traveling salesman problem: An efficient simulation algorithm. *Journal of Optimization Theory and Applications, 45*, 41–51.
7. Aharoni, T. A. (2001). *Introduction to the theory of ferromagnetism*. Oxford: Clarendon Press.
8. Bertotti, G. (1996). Energetic and thermodynamic aspects of hysteresis. *Physical Review Letters, 76*, 1739–1742.

9. Neel, L. (1950). Theorie du trainage magnetique des substances massives dans le domaine de Rayleigh. *Journal De Physique Et Le Radium, 11*, 49–61.
10. Bertotti, G., & Mayergoyz, I. (2005). *The science of hysteresis*. New York: Elsevier.
11. Néel, L. (1950). Theorie Du Trainage Magnetique Des Substances Massives Dans Le Domaine De Rayleigh. *Journal de Physique et le Radium, 11*, 49–61.
12. Weller, D., & Moser, A. (1999). Thermal effect limits in ultrahigh-density magnetic recording. *IEEE Transactions on Magnetics, 35*, 4423–4439.
13. Bertotti, G., et al. (2001). Hysteresis in magnetic materials: the role of structural disorder, thermal relaxation, and dynamic effects. *Journal of Magnetism and Magnetic Materials, 226*, 1206–1212.
14. Mitchler, P. D., et al. (1999). Interactions and thermal effects in systems of fine particles: A Preisach analysis of CrO_2 audio tape and magnetoferritin. *IEEE Transactions on Magnetics, 35*, 2029–2042.
15. Roshko, R. M., & Viddal, C. A. (2005). Non-Arrhenius relaxation effects in collections of two-level subsystems. *Physical Review B, 72*, 184422.
16. Torre, E. D. (2000). *Magnetic hysteresis*. New York: John Wiley & Sons.
17. Hodgdon, M. L. (1988). Applications of a theory of ferromagnetic hysteresis. *IEEE Transactions on Magnetics, 24*, 218–221.
18. Krzysztof, C. (2009). Modelling of dynamic hysteresis loops using the Jiles–Atherton approach. *Math Comput Model Dyn Syst, 15*, 95–105.
19. Mayergoyz, I. D., & Korman, C. E. (1991). Preisach model with stochastic input as a model for viscosity. *Journal of Applied Physics, 69*, 2128–2134.
20. Mayergoyz, I. D., & Korman, C. E. (1991). On a new approach to the modeling of viscosity in hysteretic systems. *IEEE Transactions on Magnetics, 27*, 4766–4768.
21. Korman, C. E., & Mayergoyz, I. D. (1994). The input dependent Preisach model with stochastic input as a model for aftereffect. *IEEE Transactions on Magnetics, 30*, 4368–4370.
22. Mayergoyz, I. D., & Korman, C. E. (1994). The Preisach model with stochastic input as a model for aftereffect. *Journal of Applied Physics, 75*, 5478–5480.
23. Korman, C. E., & Mayergoyz, I. D. (1996). Preisach model driven by stochastic inputs as a model for aftereffect. *IEEE Transactions on Magnetics, 32*, 4204–4209.
24. Korman, C. E., & Mayergoyz, I. D. (1997). Review of Preisach type models driven by stochastic inputs as a model for after-effect. *Physica B, 233*, 381–389.
25. Néel, L. (1949). Théorie du traînage magnétique des ferromagnétiques en grains fins avec applications aux terres cuites. *Annals of Géophysics, 5*, 99–136.
26. Andrei, P., & Stancu, A. (2006, April 15). Monte Carlo analysis of magnetic aftereffect phenomena. *Journal of Applied Physics, 99*, 08D701.
27. Mayergoyz, I. (2003). *Mathematical models of hysteresis and their applications: Electromagnetism*. Amsterdam: Academic Press.
28. Mayergoyz, I., & Dimian, M. (2003). Analysis of spectral noise density of hysteretic systems driven by stochastic processes. *Journal of Applied Physics, 93*, 6826–6828.
29. Dimian, M., & Mayergoyz, I. D. (2004). Spectral noise density of the Preisach model. *IEEE Transactions on Magnetics, 40*, 2134–2136.
30. Dimian, M., & Mayergoyz, I. D. (2004). Spectral density analysis of nonlinear hysteretic systems. *Physical Review E, 70*, 046124.
31. Adedoyin, A. (2010). Analysis of aftereffect phenomena and noise spectral properties of magnetic hysteretic systems using phenomenological models of hysteresis. *Ph.D. Thesis, Florida State University*.
32. Adedoyin, A., & Andrei, P. (2008). Data collapse and viscosity in three-dimensional magnetic hysteresis modeling. *IEEE Transactions on Magnetics, 44*, 3165–3168.
33. Mayergoyz, I. D., et al. (1999). Scaling and data collapse in magnetic viscosity. *Journal of Applied Physics, 85*, 4358–4360.

34. Mayergoyz, I. D., et al. (2000). Scaling and data collapse in magnetic viscosity (creep) of superconductors. *IEEE Transactions on Magnetics, 36*, 3208–3210.
35. Andrei, P., & Adedoyin, A. (2008, Apr 1). Phenomenological vector models of hysteresis driven by random fluctuation fields. *Journal of Applied Physics, 103*, 07D913.
36. Stancu, A., et al. (2006, April 15) Magnetic characterization of samples using first- and second-order reversal curve diagrams. *Journal of Applied Physics, 99*, 08D702.

Chapter 5
Noise Spectral Density of Hysteretic Systems

5.1 Spectral Density of Bistable Hysteretic Systems with Diffusion Input

In this section, closed form expressions for the spectral densities of bistable hysteretic systems driven by diffusion inputs are found by analytical means using the theory of stochastic processes on graphs. In the particular case of Ornstein-Uhlenbeck (OU) input, the output spectra are explicitly computed and analyzed, discussing the influence of input drift and diffusion coefficients, as well as of the rectangular loop width on the output spectra characteristics. The spectrum of bistable hysteretic system driven by colored noise is analyzed by numerical means using the Monte-Carlo method presented in Chap. 2. Since complex hysteretic nonlinearities with stochastic input can be described through Preisach formalism as weighted superposition of stochastically driven rectangular loop operators, this analysis is also useful for better understanding of spectra in complex hysteretic systems discussed in the next sections.

5.1.1 Statement of the Problem

Consider the bistable system with hysteresis represented in Fig. 5.1 that can be mathematically described by the following input–output relation:

$$I_{\beta\alpha}(t) = \hat{\gamma}_{\beta\alpha}X(t) = \begin{cases} 1, & \text{if } X(t) > \alpha, \\ -1, & \text{if } X(t) < \beta, \\ 1, & \text{if } X(t) \in (\beta, \alpha) \text{ and } X(t_-) = \alpha, \\ -1, & \text{if } X(t) \in (\beta, \alpha) \text{ and } X(t_-) = \beta, \end{cases} \tag{5.1}$$

with t_- is the value of time at which the last threshold (α or β) was attained.

The noise might have different characteristics when the system is in one state compared to the other (the transition may involve changes leading to different

M. Dimian and P. Andrei, *Noise-Driven Phenomena in Hysteretic Systems*, Signals and Communication Technology 218, DOI: 10.1007/978-1-4614-1374-5_5, © Springer Science+Business Media New York 2014

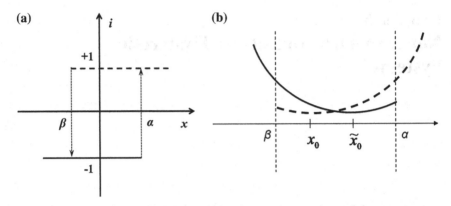

Fig. 5.1 a The input–output (x, i) diagram of a bistable system characterized by a rectangular loop; **b** Potential wells for the Brownian motion representing the noise characterization for the two metastable states in the case of OU input

internal noise characteristics). Thus, the input process $X(t)$ is assumed to be described by the Itô stochastic differential equations:

$$dX(t) = b_{\pm 1}(X(t))dt + \sigma_{\pm 1}(X(t))dW(t) \tag{5.2}$$

where $W(t)$ is the Wiener process, while $b_{\pm}$ and $\sigma_{\pm}$ are the drift and diffusion coefficients characterizing the process in $+1$ and -1, respectively.

In the particular case of Ornstein-Uhlenbeck (OU) processes, we considered $b_{+1}(x) = -b(x - x_0)$, $b_{-1}(x) = -b(x - \tilde{x}_0)$, and $\sigma_{\pm 1}(x) = \sigma$. Since the OU process can be interpreted as a Brownian motion in a parabolic potential, the noise in state $+1$ can be related to the potential represented by interrupted lines in Fig. 5.1b while the noise in state -1 can be related to the potential represented by the continuous line in Fig. 5.1b.

The autocorrelation function of the output process is:

$$C_I(\tau) = E\{I(\tau) \cdot I(0)\} = \sum_{i_\tau = \pm 1} \sum_{i_0 = \pm 1} i_\tau \cdot i_0 \cdot \rho(i_\tau, i_0) \tag{5.3}$$

where $E\{...\}$ denotes the expected value, while $\rho(i_\tau, i_0)$ is the joint probability density function. The latter is usually found from the product of the transition probability function $\rho(i_\tau | i_0)$ and the stationary probability distribution $\rho_s(i_0)$ for the given process:

$$\rho(i_\tau, i_0) = \rho(i_\tau | i_0)\rho_s(i_0) \tag{5.4}$$

According to Wiener-Khinchine theorem [1], the output spectral density can be expressed as the Fourier transform of the autocorrelation function:

$$S_i(\omega) = 2\text{Re}\left\{ \int_0^\infty C_i(\tau)e^{-j\omega\tau}d\tau \right\} \tag{5.5}$$

Although the computation of the spectral density might seem straightforward from the above presentation of the problem there is a fundamental difficulty in finding the autocorrelation function in Eq. (5.3): the hysteretic systems, even in their simplest forms, are memory dependent, and consequently, the output processes are non-Markovian. As a result, the classical approach for the calculation of autocorrelation function involves the Chapman-Kolmogorov equation for the transition probability function, which is not available for non-Markovian processes. Here, the mathematical theory of diffusion processes defined on graphs [2–4] introduced in Sect. 2.2 is used to overcome these difficulties. The presentation follows the line of articles [5–8] published by our group.

5.1.2 Embedding Output Process into a Markovian Process Defined on Graph

One can notice for a bistable hysteretic system that the joint specification of current values of input and output leads to a two dimensional stochastic process that has no memory dependence. As a result, the non-Markovian output process $I(t)$ of this system can be embedded into a two-component stochastic process $\mathbf{Z}(t) = (I(t), X(t))$ that is a Markovian process defined on graph Z shown in Fig. 5.2. According to the theory of stochastic processes on a graph, the transition probability function $\rho(\mathbf{z}_t|\mathbf{z}_0)$ for the process $\mathbf{Z}(t)$ satisfies the following forward Kolmogorov equation:

$$\frac{\partial \rho(\mathbf{z}_t|\mathbf{z}_0)}{\partial t} + \hat{L}_x^n \rho(\mathbf{z}_t|\mathbf{z}_0) = 0 \tag{5.6}$$

where $\hat{L}_x^n$ is the second order differential operator associated to the input noise process for each edge E_n of the graph. For the diffusion process (5.2) this operator has the following expression:

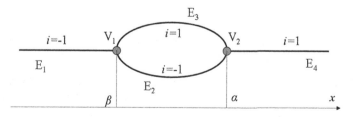

Fig. 5.2 The graph Z on which the diffusion process $\mathbf{Z}(t)$ is defined

$$\left(\hat{L}_x^n f\right)(x) = -\frac{1}{2}\frac{\partial^2}{\partial x^2}\left[\sigma_n^2(x)f(x)\right] + \frac{\partial}{\partial x}\left[b_n(x)f(x)\right] \tag{5.7}$$

The initial condition for the transition probability function has a δ distribution concentrated at z_0. In addition, the solution should decay to zero when x approaches infinity and should satisfy certain boundary conditions at the graph vertices V_1 $(x = \beta)$ and V_2 $(x = \alpha)$. These vertex boundary conditions characterize the behavior of process $Z(t)$ at the interior vertices, relating the transition probability functions that corresponds to different edges connected to a specific vertex. According to the theory of Markovian processes on graphs presented in Refs. [2–9] and summarized in Sect. 2.2 of this book, these gluing relations ensure a well-defined Markovian process on the entire graph and depend on the time spent in the vertex under consideration by the process and the probabilities that the process will "move" from the vertex along the edges connected to it.

Since our process has no delay on the vertices and there is zero probability to move from vertex V_1 along edge E_3, while random motion along the edges E_1 and E_2 are equally probable, we arrive at the following vertex boundary conditions for the process $Z(t)$ at vertex V_1 $(x = \beta)$:

$$\rho(z_t = (-1, \beta^-)|z_0) = \rho(z_t = (-1, \beta^+)|z_0), \ \rho(z_t = (1, \beta^+)|z_0) = 0,$$

$$\frac{\partial\rho}{\partial x}(z_t = (-1, \beta^-)|z_0) = \frac{\partial\rho}{\partial x}(z_t = (-1, \beta^+)|z_0) + \frac{\partial\rho}{\partial x}(z_t = (1, \beta^+)|z_0) \tag{5.8}$$

Here β^+ and β^- account for the right and left limits of the function, respectively.

In other words, these vertex boundary conditions express the continuity of the transition probability function when the move from one edge to another happens without switching the output value i and zero boundary condition is imposed on the third edge connected to that vertex. In addition, the probability current must be conserved at each vertex. Analogous boundary conditions are derived for vertex V_2 $(x = \alpha)$. In the particular case of the OU process, similar initial-boundary-value problems for the transition probability function have been postulated in [10, 11]. Here, the initial-boundary-value problem is defined on a graph and it was derived based on the theory of diffusion processes on graphs introduced in Sect. 2.2.

In conclusion, the transition probability function for Markovian process $Z(t)$ is completely defined as the solution of the initial boundary value stated above. Consequently, it can be used to compute the correlation matrix for Markovian process $Z(t)$:

$$C_Z(\tau) = \int\limits_{-\infty}^{\infty}\int\limits_{-\infty}^{\infty}\sum_{i_\tau,i_0} z_\tau^T z_0 \rho(z_\tau|z_0)\rho_s(z_0)dx_\tau dx_0 \tag{5.9}$$

where z^T denotes the transpose of vector z, while ρ_s is the stationary distribution of process $Z(t)$ satisfying the time-independent boundary value problem corresponding to the initial-boundary value problem stated above. Explicitly, it is the solution of the following set of equations:

$$\frac{1}{2}\frac{\partial^2}{\partial x^2}\left[\sigma_n^2(x)\rho_s(i,x)\right] - \frac{\partial}{\partial x}[b_n(x)\rho_s(i,x)] = 0 \tag{5.10}$$

defined on each edge E_n of the graph Z which satisfies the time independent vertex boundary conditions at vertex V_1:

$$\rho_s\left(-1,\beta^+\right) = \rho_s(-1,\beta^-), \ \rho_s\left(1,\beta^+\right) = 0, \tag{5.11}$$

similar V_2 conditions, and it decays to 0 at infinity.

Auto-correlation function $C_I(\tau)$ of the output process for a bistable system with hysteresis can be now seen and computed as the first element of the correlation matrix (5.9) for Markovian process $Z(t)$. Therefore, the fundamental difficulty related to the non-Markovian property of output process $I(t)$ was circumvented by embedding the process into two-component Markovian process $Z(t)$ defined on graph Z. We have now a well-defined path to compute the autocorrelation function of the output process and its spectral density. Since the complexity of these calculations is relatively high, several techniques are next used to reduce this complexity and to derive a closed form expression for the spectral density.

5.1.3 Closed Form Expression for Output Spectral Density

The spectral density for output process $I(t)$ is the first element of the spectral density matrix of process $Z(t)$, which is the Fourier transform of the correlation matrix given in (5.9). Once the solutions for the transition probability functions and stationary distribution are found from initial boundary value problem (5.6–5.8) and boundary value problem (5.10, 5.11), respectively, the correlation matrix (5.9) can be computed and, by taking its Fourier transform, one can compute the spectral density. However, this computation can be significantly simplified by the introduction of an auxiliary function:

$$g(\mathbf{z}, t) = \int_{-\infty}^{\infty} \sum_{i_0} i_0 \, \rho(\mathbf{z}_t = \mathbf{z}|\mathbf{z}_0)\rho_s(\mathbf{z}_0) \, dx_0 \tag{5.12}$$

and its half-line Fourier transform:

$$G(\mathbf{z}, \omega) = \int_0^{\infty} g(\mathbf{z}, \tau)e^{-j\omega\tau}d\tau \tag{5.13}$$

By using the previous definitions and relations, it can be proven that the latter function is the solution of the following equation:

$$jw\, G(\mathbf{z}, \omega) + L_x^n G(\mathbf{z}, \omega) = i \rho_s(\mathbf{z}) \tag{5.14}$$

subject to certain vertex boundary conditions. The spectral density for the output process $I(t)$ can be directly expressed in terms of $G(\mathbf{z}, \omega)$:

$$S(\omega) = 2\mathrm{Re}\left\{ \int_{-\infty}^{\infty} \sum_{i=\pm 1} i\, G(\mathbf{z}, \omega)\, dx \right\} \tag{5.15}$$

Moreover, since $(i/j\omega)\rho_s(\mathbf{z})$ is a particular solution of inhomogeneous Eq. (5.14), the function defined as:

$$G^0(\mathbf{z}, \omega) = G(\mathbf{z}, \omega) - \frac{i}{j\omega}\rho_s(\mathbf{z}) \tag{5.16}$$

is the solution of the homogeneous equation:

$$jw\, G^0(\mathbf{z}, \omega) + L_x^n G^0(\mathbf{z}, \omega) = 0 \tag{5.17}$$

subject to certain vertex boundary conditions. The particular solution of Eq. (5.14) stated above has only an imaginary part, and consequently, it does not contribute to the spectral density, which can be then written in the following form:

$$S(\omega) = 2\mathrm{Re}\left\{ \int_{-\infty}^{\infty} \sum_{i=\pm 1} i\, G^0(\mathbf{z}, \omega)\, dx \right\} \tag{5.18}$$

$G^0(\mathbf{z}, t)$ can be expressed in terms of its spatial derivatives by using Eqs. (5.18) and (5.7), fact that leads to the compensation the integral in Eq. (5.18) by the spatial derivative. As a result, the computation of the spectral density is now reduced to finding the solutions of two boundary value problems, one for stationary probability density ρ_s and one for function G^0, and calculating several derivatives of these solutions at vertex points.

For example, in the case of the OU process, the output spectral density can be written as follows:

$$S_{\alpha\beta}(\omega) = \frac{2\sigma^2}{\omega}\left\{ \left[\frac{d\rho_s}{dx}(1, \beta^+) + \frac{d\rho_s}{dx}(-1, \alpha^-) \right] \right.$$
$$\left. - \mathrm{Im}\left[\frac{dG^0}{dx}(1, \beta^+, \omega) + \frac{dG^0}{dx}(1, \alpha^+, \omega) - \frac{dG^0}{dx}(1, \alpha^-, \omega) \right] \right\}. \tag{5.19}$$

The corresponding boundary value problems can be explicitly integrated leading to analytical solutions in terms of the Gaussian functions and integrals, and the parabolic cylinder functions [12].

As a test for this method, let us consider a hard limiter (HL) system, which corresponds to the limit case when $\alpha = \beta = 0$ and, consequently it is described by the following step function:

$$I_{00}(t) = \hat{\gamma}_{00}X(t) = \begin{cases} 1, & \text{if } X(t) \geq 0 \\ -1, & \text{if } X(t) < 0 \end{cases} \tag{5.20}$$

and an Ornstein-Uhlenbeck input process defined by the following SDE:

$$dX(t) = -bX(t)dt + \sigma X(t)dW(t) \tag{5.21}$$

According to the definition of autocorrelation function for the output process:

$$C_{HL}(\tau) = E\{I_{00}(t+\tau) \cdot I_{00}(t)\} = P\{X(t+\tau)X(t) > 0\} - P\{X(t+\tau)X(t) < 0\} \tag{5.22}$$

where the probability of a negative product can be expressed as:

$$P\{X(t+\tau)X(t) < 0\} = \int_0^\infty \int_{-\infty}^0 p(x_{t+\tau}, x_t)dx_{t+\tau}dx_t + \int_{-\infty}^0 \int_0^\infty p(x_{t+\tau}, x_t)dx_{t+\tau}dx_t \tag{5.23}$$

while the probability of the positive product is simply $1 - P\{X(t+\tau)X(t) < 0\}$.

As it was proved in 2.1.5, the stationary correlation function for the OU process has expression (2.34), so the stationary joint distribution is:

$$p(x_{t+\tau}, x_t) = \frac{b}{2\pi\sigma^2\sqrt{1 - e^{-2b\tau}}} \exp\left(-\frac{b}{2\sigma^2(1 - e^{-2bt})}(x_{t+\tau}^2 + x_t^2 - 2e^{-bt}x_{t+\tau}x_\tau)\right) \tag{5.24}$$

By plugging $p(x_{t+\tau}, x_t)$ into formula (5.23) and by using the properties of Gaussian integrals, one arrives at following formula for the probability of having a negative product:

$$P\{X(t+\tau)X(t) < 0\} = \frac{1}{2} - \frac{1}{\pi}\arcsin(e^{-b\tau}) \tag{5.25}$$

and a complementary formula for the probability of a positive product:

$$P\{X(t+\tau)X(t) > 0\} = 1 - P\{X(t+\tau)X(t) < 0\} = \frac{1}{2} + \frac{1}{\pi}\arcsin(e^{-b\tau}) \tag{5.26}$$

By plugging the last two expressions into formula (5.22), the autocorrelation of the hard-limiter system is found to be:

$$C_{HL}(\tau) = \frac{2}{\pi}\arcsin(e^{-bt}) \tag{5.27}$$

This result can be traced back to the work of van Vleck [13, 14] and it is often known as "arcsine law". In conclusion, the output spectral density for the hard-limiter system (5.20) driven by the Ornstein-Uhlenbeck process (5.21) has the following analytical expression:

Fig. 5.3 Output spectral densities of the rectangular loop ($\alpha = -\beta = 0.01$; plotted with *symbols*) and the hard limiter system ($\alpha = \beta = 0$; plotted with *lines*) driven by symmetric Ornstein-Uhlenbeck type inputs ($x_0 = \tilde{x}_0$) for selected values of the drift coefficient $b = 0.5$, 1, and 3. (© 2008 NANO, [7])

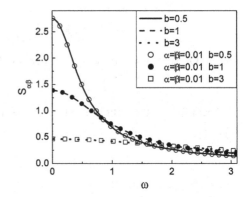

$$S_{HL}(\omega) = \frac{4}{\pi} \int\limits_{0}^{\infty} \arcsin\left(e^{-bt}\right) \cos(\omega t)dt \qquad (5.28)$$

It is expected the output spectral density for a rectangular loop should approach the output spectral density of a HL system when thresholds α and β tend to zero. Consequently, formula (5.19) can be tested in this limit case against the classical formula (5.28). The results of this comparison featured an excellent agreement between the two approaches, as it is also apparent from Fig. 5.3. Let us notice that the diffusion coefficient of the input process does not influence the spectral density for the hard-limiter system.

5.1.4 Spectral Analysis of a Bistable Hysteretic System

Next, we examine the influence of the input parameters and system characteristics on the spectral density $S_{\alpha\beta}(\omega)$ of the output of a rectangular loop based on Refs. [6–9]. Besides the interest in its own right, this analysis will be also useful for the understanding of the spectral density of Preisach systems. In Fig. 5.4, the dependence of the spectral noise density on the loop width is presented. For narrow loops, the spectral noise density is similar to the one of a step operator (hard limiter system) where the region of the white noise is connected to the region of $1/f^2$ noise through an intermediate region of $1/f$ behavior (the frequency $f = \omega/2\pi$). This intermediate frequency region is reduced as the loop is broadened, and the variations of the loop width lead mostly to self-similar transformations of the spectral noise density graph. Another interesting observation that emerges from this analysis is related to the transformation of the spectral band. It is known that memoryless nonlinearities broaden spectral bands. However, memory effects may lead to opposite results as shown in Fig. 5.4.

By analyzing the formula for $S_{\alpha\alpha}(\omega)$ and the related boundary-value problems, the following scaling property can be derived:

Fig. 5.4 Spectral density $S_{\alpha\alpha}$ of a rectangular loop for various widths of the loop α plotted in a *log–log* scale. The Ornstein-Uhlenbeck input parameters are: $b = \sigma=1$, $x_0 = \tilde{x}_0 = 0$. (© 2004 APS, [6])

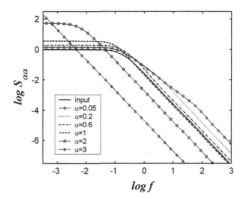

$$S_{\alpha\alpha}((x_0, b, \sigma), \omega) = \left(\frac{\alpha}{\sigma}\right)^2 S_{11}\left(\left(\frac{x_0}{\alpha}, b\left(\frac{\alpha}{\sigma}\right)^2, 1\right), \omega\left(\frac{\alpha}{\sigma}\right)^2\right) \qquad (5.29)$$

The advantage provided by formula (5.29) is that the computation of the output spectral density for a symmetric rectangular loop with OU inputs is reduced to the computation of the spectral density S_{11}.

The effect of the input stationary average x_0 can be seen from Fig. 5.5. It is apparent that when x_0 is increased, output signals "stabilize" around $+1$ and, consequently, the spectral noise density is diminished. The influence of the drift coefficient b (or its inverse that represents the correlation time of the input process) on the output spectral density is represented in Fig. 5.6. Thus, sample noise spectra for a hard limiter system ($\alpha = \beta = 0$) are plotted in Fig. 5.6a, while sample spectra for a rectangular loop with $\alpha = -\beta = 1$ are plotted in Fig. 5.6b. In the insets of the two figures, the level of flat spectrum region is plotted against b. As it is apparent from the two figures and their insets, the influence of input noise temporal correlation on the output spectra is quite different when the loop width is negligible with respect to noise strength than in the case when the two are comparable. When the noise strength is dominant ($\alpha - \beta \ll \sigma$), the system behaves as a hard limiter system: The bandwidth of the output spectrum is narrowing, and the

Fig. 5.5 Spectral density S_{11} for various values of the input average value x_0 ($b = \sigma = 1$). (© 2004 APS, [6])

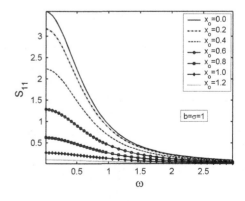

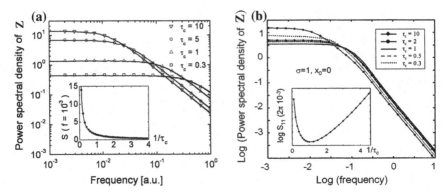

Fig. 5.6 (**a**) Power spectral density for a hard limiter system for selected values of correlation time $\tau_c = 1/b$ plotted in a *log–log* scale ($\sigma = 1$, $x_0 = 0$); in the inset, the level of flat spectrum region is plotted against b; (**b**) Spectral density S_{11} for selected values of the correlation time $\tau_c = 1/b$ plotted in a *log–log* scale ($\sigma = 1$, $x_0 = 0$); in the inset, the level of flat spectrum region against b. (© 2010 IEEE, [8])

level of flat spectrum region is increasing as the correlation time increases (see the inset of Fig. 5.6a). When the memory property becomes prominent, the monotonic behavior presented above is no longer valid, and some extrema for the flat spectrum level and bandwidth appear at some specific correlation time (see the inset of Fig. 5.6b).

These analytical results are also providing the opportunity to test the Monte-Carlo approach to modeling and simulation noise induced phenomena in hysteretic systems presented/and tested for thermal relaxation phenomena in the previous chapter. A very good agreement is observed between the numerical simulations and the analytical results for the spectral analysis of rectangular loops driven by Ornstein-Uhlenbeck noise, which demonstrates the reliability and accuracy of Monte-Carlo technique developed for stochastic hysteretic system.

By these numerical means implemented in HysterSoft©, the spectral analysis of rectangular loop driven by colored noise can be performed. When the noise spectrum increases with frequency as f^2, so-called *violet* noise, the output spectrum stays almost constant for most of the frequency interval, except for high-frequency region where it features an increase slightly higher than a linear dependence of f. When the noise spectrum increases with frequency as f, so-called *blue* noise, a similar behavior is observed for low-frequency region, while a slight spectrum increase is observed for high-frequency region. The simulations performed for various input power-law spectrum f^λ, $\lambda > 0$, lead to the conclusion that the corresponding output spectra feature a flat region for low-frequency region and a power law with an exponent slightly different than for high frequency.

The behavior corresponding to a power-law input spectrum f^λ, $\lambda < 0$, is significantly different than the one described above. When the noise spectrum is proportional to $1/f$, coined as *pink* noise, the output spectrum follows a similar pattern for low frequency, with an exponent slightly higher than 1, while a $1/f^2$

behavior is present for high frequencies. When the noise spectrum is proportional to $1/f^2$, known as *Brownian* noise, the output spectrum follows a similar behavior, except that the exponent is slightly higher than 2 for low-frequency region. The decay in the output spectra for high frequency was common to all power-law input spectra with considered in our analysis.

Sample of these simulations for output spectrum of rectangular loop $\hat{\gamma}_{-1,1}$ driven by colored Gaussian noise are shown in Fig. 5.7. By using HysterSoft© the reader can generate a wide variety of colored noises and analyze their spectral transformation by a general bistable system with hysteretic rectangular loop.

Next, let us consider that noise has different characteristics in one state of the system than in the other. As a case study, we take a noise described by an OU

Fig. 5.7 Noise input spectrum and the corresponding output spectrum for rectangular loop $\hat{\gamma}_{-1,1}$ in the case of (**a**) *pink* Gaussian noise input ($1/f$) and *Brownian* noise input ($1/f^2$); (**b**) *blue* Gaussian noise input (f) and *violet* Gaussian noise input (f^2)

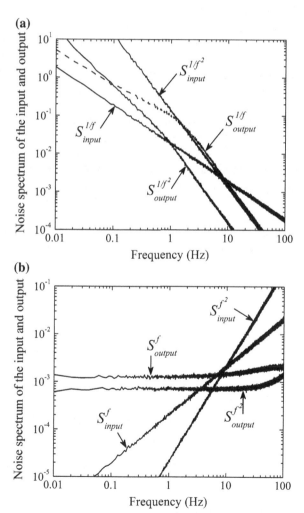

process with $b_{+1}(x) = -b(x - x_0)$ in $+1$ state and by an OU process with $b_{-1}(x) = -b(x - \tilde{x}_0)$ in -1 state, while diffusion coefficient is the same in both states. As mentioned above, noise in state $+1$ can be interpreted as a Brownian motion in a parabolic potential represented by the interrupted line in Fig. 5.1b while the noise in state -1 can be interpreted as a Brownian motion in a parabolic potential represented by the continuous line in Fig. 5.1b. In the case of symmetric noise ($x_0 = \tilde{x}_0$), the monotonic behavior of the output spectral density with respect to the frequency is a common feature of bistable hysteretic system. The symmetry breaking ($x_0 \neq \tilde{x}_0$) can lead to non-monotonic behavior and more precisely to the appearance of a maximum in the output spectra, as can be observed from Fig. 5.8 obtained by using Eq. (5.19). That can be related to the manifestation of an almost regular behavior of the system output, so pure noise input can lead to almost periodic sequences of -1 and 1. This noise induced phenomena is known as *coherence resonance* [11, 15].

In conclusion, output power spectral density of bistable hysteretic systems with diffusion input has been found by analytical means by using the theory of stochastic processes on graphs. In the particular case of OU input, the output spectra have been explicitly computed and analyzed, discussing the influence of input drift and diffusion coefficients, as well as of the rectangular loop width on the output spectra characteristics. While it is mostly experienced as a disruptive effect, noise can also have a constructive role, activating a resonance response of the system. It was proven that certain bistable hysteretic systems driven by "state-dependent" noise inputs manifest of an almost regular behavior of the system output. The spectrum of bistable hysteretic system driven by colored noise has been analyzed by numerical means using the Monte-Carlo method presented in Chap. 2. Since complex hysteretic nonlinearities with stochastic input can be described through Preisach formalism as weighted superposition of stochastically driven rectangular loop operators, this analysis is also useful for better understanding of spectra in complex hysteretic systems such as the ones discussed in the next sections.

Fig. 5.8 Output spectral densities of the rectangular loop ($\alpha = -\beta = 0.5$) driven by asymmetric Ornstein-Uhlenbeck type inputs for selected values of the diffusion coefficient σ. (© 2008 NANO, [7])

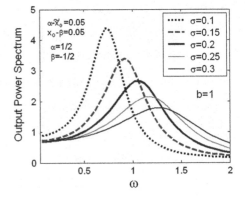

5.2 Spectral Density of Symmetric Preisach Systems with Diffusion Input: Analytical Approach

In this section, closed form expressions for the spectral densities of symmetric Preisach hysteretic systems driven by diffusion inputs are found by analytical means. The theory of stochastic processes on graphs is used to circumvent the difficulties related to the non-Markovian property of the output of hysteretic systems, while the explicit calculations are appreciably simplified by the introduction of the "effective" distribution function. The implementation of the method for the case of Ornstein-Uhlenbeck process is presented in detail and general qualitative features of these spectral densities are examined. Due to the universality of the Preisach model, this approach can be used to describe hysteresis nonlinearities of various physical origins.

5.2.1 Statement of the Problem

Consider complex hysteretic nonlinearities that can be modeled through the Preisach formalism (see Sect. 1.2) as weighted superposition of rectangular loops. For many hysteretic systems (especially magnetic materials), the Preisach distribution is narrowly peaked around the diagonal line $\alpha = -\beta$ and consequently, it can be approximated by $\mu(\alpha)\delta(\alpha + \beta)$. For these materials, the symmetric Preisach model is constructed as a weighted superposition of symmetric rectangular loops $\hat{\gamma}_\alpha = \hat{\gamma}_{\alpha(-\alpha)}$ with the weight function $\mu(\alpha)$ which will be considered Preisach distribution for that symmetric system. Thus the symmetric Preisach model takes the following form:

$$y(t) = \int_0^{\alpha_0} \hat{\gamma}_\alpha x(t)\mu(\alpha)d\alpha = \int_0^{\alpha_0} i_\alpha(t)\mu(\alpha)d\alpha \qquad (5.30)$$

where:

$$i_\alpha(t) = \hat{\gamma}_\alpha x(t) = \begin{cases} 1, & \text{if } x(t) > \alpha, \\ -1, & \text{if } x(t) < -\alpha, \\ 1, & \text{if } x(t) \in (-\alpha, \alpha) \text{ and } x(t_-) = \alpha, \\ -1, & \text{if } x(t) \in (-\alpha, \alpha) \text{ and } x(t_-) = -\alpha, \end{cases} \qquad (5.31)$$

with t_- is the value of time at which the last threshold (α or $-\alpha$) was attained.

The input process $x(t)$ is assumed to be described by the Itô stochastic differential equation:

$$dX(t) = b(X(t))dt + \sigma(X(t))dW(t) \qquad (5.32)$$

where $W(t)$ is the Wiener process, while b and σ are the drift and diffusion coefficients, respectively. The stochastic nature of the input leads to random switchings of the rectangular loop operators $\hat{\gamma}_\alpha$ and, therefore, the output of the Preisach model is a stochastic process as well, denoted by $Y(t)$.

The autocorrelation function of the output process $Y(t)$ is:

$$C_Y(\tau) = E\{Y(\tau)Y(0)\} = \int_0^{\alpha_0} \int_0^{\alpha_0} E\{\hat{\gamma}_\beta X(\tau)\hat{\gamma}_\alpha X(0)\}\mu(\beta)\mu(\alpha)d\beta d\alpha, \qquad (5.33)$$

Thus, we can express the autocorrelation function as a weighted superposition of cross-correlation functions $C_{\beta\alpha}(\tau)$ of two-dimensional processes $(I_\beta(t), I_\alpha(t))$, representing the outputs of two symmetric rectangular loops:

$$C_Y(\tau) = \int_0^{\alpha_0} \int_0^{\alpha_0} C_{\beta\alpha}(\tau)\mu(\beta)\mu(\alpha)d\beta d\alpha, \qquad (5.34)$$

Cross-correlation functions $C_{\beta\alpha}(\tau)$ are not even functions, but $C_{\beta\alpha}(-\tau) = C_{\alpha\beta}(\tau)$ and consequently, the correlation function of the Preisach system $C_Y(\tau)$ is even.

According to the Wiener-Kinchine theorem [10], the process's spectral density is the Fourier Transform of the autocorrelation function. Because we deal with an even correlation function, the spectral density of the output process can be expressed as:

$$S_Y(\omega) = 2\text{Re}\left\{\int_0^\infty C_Y(\tau)e^{-j\omega\tau}d\tau\right\} = \int_0^{\alpha_0}\int_0^{\alpha_0} S_{\beta\alpha}(\omega)\mu(\beta)\mu(\alpha)d\beta d\alpha, \qquad (5.35)$$

where $S_{\beta\alpha}(\omega)$ is the "cross-spectral density" for the two-dimensional process $(I_\beta(t), I_\alpha(t))$ and it is related to the cross-correlation function $C_{\beta\alpha}(\tau)$ as follows:

$$S_{\beta\alpha}(\omega) = 2\text{Re}\left\{\int_0^\infty C_{\beta\alpha}(\tau)e^{-j\omega\tau}d\tau\right\}. \qquad (5.36)$$

5.2.2 Calculation Method for the Output Correlation Function Using Markovian Processes on Graphs

The Preisach model describes hysteresis nonlinearities with non-local memories. For this reason, the output process $Y(t)$ cannot be embedded as a component of some finite-dimensional Markov process. However, the previous expression shows that this spectral density can be expressed as a weighted superposition of spectral densities for much simpler processes ($I_\beta(t)$, $I_\alpha(t)$). These processes are still non-Markov, but they can be embedded in higher dimensional Markov processes.

In order to compute $S_{\beta\alpha}(\omega)$, let us consider the three component process $\mathbf{Z}(t) = \big(I_\beta(t), I_\alpha(t), X(t)\big)$. Because the rectangular loop operators describe hysteresis with local memory, the joint specification of current values of input and output uniquely define the states of this hysteresis. As a result, $\mathbf{Z}(t)$ is a Markovian process. In addition, only certain combinations of $I_\beta(t)$, $I_\alpha(t)$ and $X(t)$ are possible, and they are presented on the graph Z shown in Fig. 5.9. The binary process $I_\beta(t)$ and $I_\alpha(t)$ assume constant values on edges of the graph Z.

Applying the theory of stochastic processes on graphs (see Sect. 2.2.2), the following initial-boundary value problem for the transition probability density function $\rho(\mathbf{z}, t | \mathbf{z}', 0)$ of the Markovian process $\mathbf{Z}(t)$ defined on the graph Z can be derived. On each edge of this graph, $\rho(\mathbf{z}, t | \mathbf{z}', 0)$ satisfies the following forward Kolmogorov equation:

$$\frac{\partial \rho(\mathbf{z}, t | \mathbf{z}', 0)}{\partial t} + L_x \rho(\mathbf{z}, t | \mathbf{z}', 0) = 0 \tag{5.37}$$

where $\hat{L}_x$ is the second order elliptic operator associated with the input diffusion process defined in (5.32) and is specified by the expression:

$$\hat{L}_x \rho = -\frac{1}{2}\frac{\partial^2}{\partial x^2}\big(\sigma^2(x)\rho\big) + \frac{\partial}{\partial x}\big(b(x)\rho\big) \tag{5.38}$$

The function $\rho(\mathbf{z}, t | \mathbf{z}', 0)$ satisfies the initial conditions:

$$\rho(\mathbf{z}, 0 | \mathbf{z}', 0) = \delta_{i_\beta i_\beta'}\,\delta_{i_\alpha i_\alpha'}\,\delta(x, x') \tag{5.39}$$

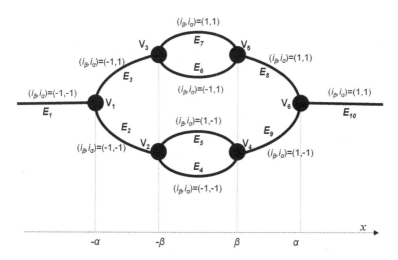

Fig. 5.9 The graph on which three component process Z is defined

and it has to decay to zero for x going to infinity. In addition, the so-called "vertex" type boundary conditions (2.73) at graph vertices have to be satisfied. These "vertex" type boundary conditions express the continuity of the transition probability density when the transition from one graph edge occurs without switching of the rectangular loop, and zero boundary condition is imposed on the third graph edge connected to this vertex. Moreover, the probability current has to be conserved at each vertex. For example, at the vertex V_1 (corresponding to $x = -\alpha$, in the case $\alpha > \beta$), these conditions are explicitly written as:

$$\rho((-1,-1,-\alpha^+),t|\mathbf{z}',0) = \rho((-1,-1,-\alpha^-),t|\mathbf{z}',0),$$
$$\rho((-1,1,-\alpha^+),t|\mathbf{z}',0) = 0,$$
$$\frac{\partial\rho}{\partial x}((-1,1,-\alpha^+),t|\mathbf{z}',0) + \frac{\partial\rho}{\partial x}((-1,-1,-\alpha^+),t|\mathbf{z}',0) = \frac{\partial\rho}{\partial x}((-1,-1,-\alpha^-),t|\mathbf{z}',0).$$

$$(5.40)$$

It is apparent that the stationary probability density of the process $\mathbf{Z}(t)$ is the solution of the following boundary value problem:

$$\begin{cases} \hat{L}_x\rho_s(\mathbf{z}) = 0 \text{ on each graph edge,} \\ \text{"vertex" boundary conditions at each graph vertex.} \end{cases} \qquad (5.41)$$

Taking into account the facts presented above, the cross-correlation function $C_{\beta\alpha}(\tau)$ can be seen as a component of the correlation matrix $C_{\mathbf{Z}}(\tau)$ for the Markov process $\mathbf{Z}(t)$:

$$C_{\mathbf{Z}}(\tau) = E\{\mathbf{Z}^T(\tau)\mathbf{Z}(0)\} = \int_{-\infty}^{\infty}\int_{-\infty}^{\infty}\sum_{i_\alpha,i_\beta}\sum_{i'_\alpha,i'_\beta}\mathbf{z}^T\mathbf{z}'\rho(\mathbf{z},\tau;\mathbf{z}',0)dxdx'$$
$$= \int_{-\infty}^{\infty}\int_{-\infty}^{\infty}\sum_{i_\alpha,i_\beta}\sum_{i'_\alpha,i'_\beta}\mathbf{z}^T\mathbf{z}'\rho(\mathbf{z},\tau|\mathbf{z}',0)\rho_s(\mathbf{z}')dxdx'$$

$$(5.42)$$

In the above formula, the sums are taken over all graph values of the (i_β, i_α) and (i'_β, i'_α), respectively. This convention is maintained throughout the book.

5.2.3 Closed Form Expression for the Output Spectral Density

To simplify the computation of the cross-correlation function, the "effective" distribution function $g(\mathbf{z}, \tau)$ is introduced:

$$g(\mathbf{z}, \tau) = \int_{-\infty}^{\infty}\sum_{i'_\alpha,i'_\beta}i'_\alpha\rho(\mathbf{z},\tau|\mathbf{z}',0)\rho_s(\mathbf{z}')dx' \qquad (5.43)$$

A similar function has been previously proposed in [16] and used in the analysis of noise in semiconductor devices.

By using Eq. (5.37) on each edge of the graph, the initial condition (5.39), and "vertex" type boundary conditions for transition probability density, as well as boundary value problem (5.41) for stationary probability density, one can derive the following initial boundary value problem for the "effective" distribution function:

$$
\begin{cases}
\frac{\partial g(\mathbf{z},\tau)}{\partial \tau} + L_x g(\mathbf{z},\tau) = 0 \text{ on each graph edge,} \\
g(\mathbf{z},0) = i_\alpha \rho_s(\mathbf{z}), \\
\lim_{x \to \pm\infty} g(\mathbf{z},\tau) = 0, \\
\text{``vertex'' boundary conditions.}
\end{cases}
\tag{5.44}
$$

Using formulas (5.42) and (5.43) the cross-correlation function $C_{\beta\alpha}(\tau)$ can be expressed by the formula:

$$
C_{\beta\alpha}(\tau) = \int_{-\infty}^{\infty} \sum_{i_\alpha,i_\beta} i_\beta g\big((i_\beta, i_\alpha, x), \tau\big) dx
\tag{5.45}
$$

Thus, in order to find the cross-correlation function $C_{\beta\alpha}(\tau)$, one has to solve first the boundary value problem (5.41) for stationary distribution $\rho_s(\mathbf{z})$, then the initial-boundary value problem (5.44) for the "effective" distribution function $g(\mathbf{z},\tau)$, and finally to compute integral (5.45). According to the Eq. (5.36), another integration has to be performed for the computation of the cross-spectral density $S_{\beta\alpha}(\omega)$. However, by introducing the one-side Fourier transform of the "effective" distribution function:

$$
G(\mathbf{z},\omega) = \int_0^{\infty} g(\mathbf{z},\tau) e^{-j\omega\tau} d\tau
\tag{5.46}
$$

the cross-spectral density $S_{\beta\alpha}(\omega)$ can be written in the form:

$$
S_{\beta\alpha}(\omega) = 2\mathrm{Re}\left\{ \int_0^{\infty} \sum_{i_\alpha,i_\beta} i_\beta G(\mathbf{z},\omega) dx \right\}
\tag{5.47}
$$

Performing the Fourier transformation of the initial-boundary-value problem (5.44), we arrive at the following boundary-value problem for $G(\mathbf{z},\omega)$:

$$
\begin{cases}
j\omega G(\mathbf{z},\omega) + L_x G(\mathbf{z},\omega) = i_\alpha \rho_s(\mathbf{z}) \text{ on each graph edge,} \\
\lim_{x \to \pm\infty} G(\mathbf{z},\omega) = 0, \\
\text{``vertex'' boundary conditions.}
\end{cases}
\tag{5.48}
$$

For example, these "vertex" boundary conditions at vertex V_1 ($x = -\alpha$) are:

$$G((-1, -1, -\alpha^+), \omega) = G((-1, -1, -\alpha^-), \omega),$$
$$G((-1, 1, -\alpha^+), \omega) = 0,$$
$$\frac{\partial G}{\partial x}((-1, 1, -\alpha^+), \omega) + \frac{\partial G}{\partial x}((-1, -1, -\alpha^+), \omega) = \frac{\partial G}{\partial x}((-1, -1, -\alpha^-), \omega)$$

$$(5.49)$$

Because the stationary probability distribution satisfies the differential equation of the boundary-value problem (5.41), function $(i_\alpha/j\omega)\rho_s(\mathbf{z})$ is (for each ω) a particular solution for the non-homogeneous differential equation in (5.48). Taking into account the linearity of operator L_x, $G(\mathbf{z}, \omega)$ can be written as:

$$G(\mathbf{z}, \omega) = G^0(\mathbf{z}, \omega) + \frac{i_\alpha}{j\omega}\rho_s(\mathbf{z}) \tag{5.50}$$

where $G^0(\mathbf{z}, \omega)$ is a solution of the corresponding homogeneous equation. Since the particular solution is purely imaginary, it does not contribute to the cross-spectral density $S_{\beta\alpha}(\omega)$. Thus,

$$S_{\beta\alpha}(\omega) = 2\text{Re}\left\{ \int_0^\infty \sum_{i_\alpha, i_\beta} i_\beta G^0(\mathbf{z}, \omega) dx \right\} \tag{5.51}$$

with $G^0(\mathbf{z}, \omega)$ satisfying the following boundary-value problem:

$$\begin{cases} j\omega G^0(\mathbf{z}, \omega) + L_x G^0(\mathbf{z}, \omega) = 0 \text{ on each graph edge,} \\ \lim_{x \to \pm\infty} G^0(\mathbf{z}, \omega) = 0, \\ \text{inhomogeneous ``vertex'' - type boundary conditions.} \end{cases} \tag{5.52}$$

Next, we describe these inhomogeneous "vertex"-type boundary conditions. First, by inspecting "vertex" boundary conditions for $G(\mathbf{z}, \omega)$ and $\rho_s(\mathbf{z})$, it can be observed that, when transition from one edge to another occurs without switching of the rectangular loops, $G(\mathbf{z}, \omega)$ and $\rho_s(\mathbf{z})$ corresponding to these edges are continuously matched and i_α does not change its value. Consequently, the corresponding $G^0(\mathbf{z}, \omega)$ is also continuously matched in this case. On the third edge connected to the vertex, zero boundary condition is valid. Until this point, inhomogeneous "vertex" boundary type conditions coincide with the previous ones. This coincidence is also maintained in the boundary conditions for derivatives at vertices V_{2-5} ($x = \pm\beta$). However, the difference appears in the conditions for derivatives at vertices V_1 and V_6 ($x = \pm\alpha$). Namely, from the boundary condition (5.49) for the derivative of $G(\mathbf{z}, \omega)$, we have:

$$\frac{\partial G^0}{\partial x}((-1,1,-\alpha^+),\omega) + \frac{1}{j\omega}\frac{\partial \rho_s}{\partial x}(-1,1,-\alpha^+) + \frac{\partial G^0}{\partial x}((-1,-1,-\alpha^+),\omega)$$

$$+\frac{(-1)}{j\omega}\frac{\partial \rho_s}{\partial x}(-1,-1,-\alpha^+) = \frac{\partial G^0}{\partial x}((-1,-1,-\alpha^-),\omega) + \frac{(-1)}{j\omega}\frac{\partial \rho_s}{\partial x}(-1,-1,-\alpha^-).$$

$$(5.53)$$

Taking into account the boundary condition for stationary probability distribution, the following boundary condition for $G^0(\mathbf{z}, \omega)$ is derived:

$$\frac{\partial G^0}{\partial x}((-1,1,-\alpha^+),\omega) + \frac{\partial G^0}{\partial x}((-1,-1,-\alpha^+),\omega)$$

$$+\frac{2}{j\omega}\frac{\partial \rho_s}{\partial x}(-1,1,-\alpha^+) = \frac{\partial G^0}{\partial x}((-1,-1,-\alpha^-),\omega). \qquad (5.54)$$

By using similar arguments, the inhomogeneous "vertex" boundary condition at the vertex V_6 is found to be:

$$\frac{\partial G^0}{\partial x}((1,1,\alpha^-),\omega) + \frac{\partial G^0}{\partial x}((1,-1,\alpha^-),\omega)$$

$$-\frac{2}{j\omega}\frac{\partial \rho_s}{\partial x}(1,-1,-\alpha^+) = \frac{\partial G^0}{\partial x}((1,1,\alpha^+),\omega). \qquad (5.55)$$

In the case $\alpha < \beta$, the boundary conditions for vertices corresponding to $x = \mp\alpha$ take the following form:

$$\frac{\partial G^0}{\partial x}((i_\beta,1,\mp\alpha^\pm),\omega) + \frac{\partial G^0}{\partial x}((i_\beta,-1,\mp\alpha^\pm),\omega)$$

$$\pm\frac{2}{j\omega}\frac{\partial \rho_s}{\partial x}(i_\beta,\pm1,\mp\alpha^\pm) = \frac{\partial G^0}{\partial x}((i_\beta,\mp1,\mp\alpha^\mp),\omega). \qquad (5.56)$$

Now, the method for the calculation of the spectral density can be summarized as the sequence of the following steps:

Step 1: Solve boundary value problem (5.41) for stationary distribution $\rho_s(\mathbf{z})$.
Step 2: Solve boundary value problem (5.52) for $G^0(\mathbf{z}, \omega)$
Step 3: Calculate cross-spectral density $S_{\beta\alpha}(\omega)$ by using formula (5.51)
Step 4: Calculate spectral density $S_Y(\omega)$ by using formula (5.35).

The following observations can simplify the implementation of the above steps:

1. For a given input, first three steps of the method are independent of Preisach function $\mu(\alpha)$. Therefore, once $S_{\beta\alpha}(\omega)$ are precomputed, they can be used for any "symmetric" Preisach system (5.30). In other words, the spectral density of a hysteretic system can be computed as a weighted superposition of cross-spectral densities $S_{\beta\alpha}(\omega)$ precomputed at the third step, with the weight being given by the Preisach function of that system.

2. As can be observed from Eq. (5.51), the cross-spectral densities $S_{\beta\alpha}(\omega)$ are expressed as linear combinations of $G^0(\mathbf{z}, \omega)$ corresponding to different edges. This indicates that it may not be necessary to find an explicit expression for $G^0(\mathbf{z}, \omega)$ on every edge, but rather their linear combinations mentioned above.

3. By using the expression (5.38) for operator $\hat{L}_x$, an important simplification can be made. From formula (5.52) follows $G^0(\mathbf{z}, \omega) = (j/\omega)\hat{L}_x G^0(\mathbf{z}, \omega)$. By substituting the later expression into formula (5.51), one can obtain:

$$S_{\beta\alpha}(\omega) = 2\text{Re}\left\{ \int_0^\infty \sum_{i_\alpha, i_\beta} i_\beta \left(\frac{j}{\omega}\right) \hat{L}_x G^0(\mathbf{z}, \omega) dx \right\}$$

$$= -\frac{2}{\omega} \text{Im}\left\{ \int_0^\infty \hat{L}_x \left(\sum_{i_\alpha, i_\beta} i_\beta G^0(\mathbf{z}, \omega) \right) dx \right\}. \tag{5.57}$$

The derivatives in the operator $\hat{L}_x$ can be integrated and this results in a simple expression for the spectral density in terms of the first derivatives of $G^0(\mathbf{z}, \omega)$ at vertex points (see, for example, the next section).

4. The boundary-value problems (5.41) and (5.52) defined on the entire graph Z, can be sequentially reduced to the boundary-value problems defined on the real line intervals which are more tractable analytically and numerically. Efficient numerical algorithms for solving these problems defined on the real line interval are three-diagonal matrix solvers described, for instance, in Ref. [17].

The above observations produce further simplifications in the method for computations of the spectral density once a specific form of the input stochastic process is given. These advantages will be further exploited in the next section where the Ornstein-Uhlenbeck process is used as a model of driving noise.

The proposed method is conceptually valid for Preisach systems with non-symmetric rectangular loops, although the complexity of calculations will be appreciably increased.

5.2.4 Example: Spectral Density of Symmetric Preisach Systems with Ornstein-Uhlenbeck Input

In this section we shall apply the method developed in the previous section to the case when the input is an Ornstein-Uhlenbeck (OU) process. As has been discussed in 2.1.5, the OU process satisfies:

$$\hat{L}_x \rho = -b \frac{\partial[(x - x_0)\rho]}{\partial x} - \frac{\sigma^2}{2} \frac{\partial^2 \rho}{\partial x^2} \tag{5.58}$$

Step 1

For the OU input process, the boundary-value problem (5.41) for the stationary distribution of the process $\mathbf{z}(t)$ defined on the graph Z from Fig. 5.9 can be solved by using the first example in Sect. 2.2.3. Thus, by adding the stationary distributions corresponding to edges E_6 and E_7 as well as to edges E_4 and E_5, we end up to the problem solved there and the results are:

$$
\begin{aligned}
\tilde{\rho}_1^{st}(x) &= \hat{\rho}^{st}(x), \quad x \in (-\infty, -\alpha)\\
\tilde{\rho}_2^{st}(x) &= \hat{\rho}^{st}(x)(1 - \phi_{-\alpha\alpha}(x)), \quad x \in (-\alpha, \alpha)\\
\tilde{\rho}_3^{st}(x) &= \hat{\rho}^{st}(x)\phi_{-\alpha\alpha}(x), \quad x \in (-\alpha, \alpha)\\
\tilde{\rho}_4^{st}(x) &= \hat{\rho}^{st}(x), \quad x \in (\alpha, \infty)
\end{aligned}
\tag{5.59}
$$

where

$$
\hat{\rho}_s(x) = \sqrt{\frac{b}{\pi\sigma^2}}e^{-b(x-x_0)^2/\sigma^2}, \quad \phi_{a_1 a_2}(x) = \frac{\int_{a_1}^{x} e^{b(y-x_0)^2/\sigma^2}}{\int_{a_1}^{a_2} e^{b(y-x_0)^2/\sigma^2}}
\tag{5.60}
$$

Similarly to the previous derivation, the components of the stationary distributions for edges E_6 and E_7 as well as for edges E_4 and E_5 can be determined leading to the following expression:

$$
\rho_s(\mathbf{z}) = \begin{cases}
\hat{\rho}_s(x) \text{ on } E_1 \text{ and} E_{10},\\
\hat{\rho}_s(x)(1 - \phi_{-\alpha\alpha}(x)) \text{ on } E_2 \text{ and } E_8,\\
\hat{\rho}_s(x)\phi_{-\alpha\alpha}(x) \text{ on } E_3 \text{ and } E_9,\\
\hat{\rho}_s(x)(1 - \phi_{-\alpha\alpha}(x))(1 - \phi_{-\beta\beta}(x)) \text{ on } E_4,\\
\hat{\rho}_s(x)(1 - \phi_{-\alpha\alpha}(x))\phi_{-\beta\beta}(x) \text{ on } E_5,\\
\hat{\rho}_s(x)\phi_{-\alpha\alpha}(x)(1 - \phi_{-\beta\beta}(x)) \text{ on } E_6,\\
\hat{\rho}_s(x)\phi_{-\alpha\alpha}(x)\phi_{-\beta\beta}(x) \text{ on } E_7,
\end{cases}
\tag{5.61}
$$

The results for the case $\alpha < \beta$ are obtained by interchanging α and β.

Step 2

Next, the boundary-value problem (5.52) defined on the graph Z is reduced to boundary-value problems defined on line intervals, which are better tractable both analytically and numerically. This procedure is very useful because it could be applied to Steps 1 and 2 of the method in the case of a general input diffusion process.

First, we formulate the boundary-value problem for $G^0(x, \omega) = \sum_{i_\alpha, i_\beta} G^0(\mathbf{y}, \omega)$, where the sum is taken over all graph edges:

$$
\begin{cases}
j\omega G^0(x, \omega) + L_x G^0(x, \omega) = 0, \ x \in (-\infty, +\infty)\setminus\{-\alpha, \alpha\},\\
\lim_{x \to \pm\infty} G^0(x, \omega) = 0,\\
\frac{\partial G^0}{\partial x}(-\alpha^-, \omega) - \frac{\partial G^0}{\partial x}(-\alpha^+, \omega) = \frac{2}{j\omega}\sqrt{\frac{b}{\pi\sigma^2}}\left(\int_{-\alpha}^{\alpha} e^{b(y-x_0)^2/\sigma^2}\right)^{-1},\\
\frac{\partial G^0}{\partial x}(\alpha^-, \omega) - \frac{\partial G^0}{\partial x}(\alpha^+, \omega) = -\frac{2}{j\omega}\sqrt{\frac{b}{\pi\sigma^2}}\left(\int_{-\alpha}^{\alpha} e^{b(y-x_0)^2/\sigma^2}\right)^{-1}.
\end{cases}
\tag{5.62}
$$

The solution of this problem coincides with the solution of problem (5.52) for edges E_1 and E_{10}. In addition, it will also help to simplify the expression for the cross-spectral density.

Second, we formulate the boundary-value problem for $G^0(1, x, \omega) = \sum_{i_\beta} G^0((1, i_\beta, x), \omega)$, where the sum is taken over "central" graph edges. In the case $\alpha < \beta$, $G^0(1, x, \omega) = \sum_{i_\alpha} G^0((i_\alpha, 1, x), \omega)$.

From formulas (5.52) and (5.61), we find:

$$\begin{cases} j\omega G^0(1, x, \omega) + L_x G^0(1, x, \omega) = 0, \ x \in (-\alpha, \alpha), \\ G^0(1, -\alpha, \omega) = 0, \\ G^0(1, \alpha, \omega) = G^0(\alpha, \omega) \end{cases} \tag{5.63}$$

for the case of $\alpha > \beta$ and

$$\begin{cases} j\omega G^0(1, x, \omega) + L_x G^0(1, x, \omega) = 0, \ x \in (-\beta, \beta) \setminus \{-\alpha, \alpha\}, \\ G^0(1, -\beta, \omega) = 0, \\ G^0(1, \beta, \omega) = G^0(\beta, \omega), \\ \frac{\partial G^0}{\partial x}(1, -\alpha^-, \omega) - \frac{\partial G^0}{\partial x}(1, -\alpha^+, \omega) = \frac{2}{j\omega} \sqrt{\frac{b}{\pi\sigma^2}} \frac{\phi_{-\beta\beta}(-\alpha)}{\int_{-\alpha}^{\alpha} e^{b(y-x_0)^2/\sigma^2}}, \\ \frac{\partial G^0}{\partial x}(1, \alpha^-, \omega) - \frac{\partial G^0}{\partial x}(1, \alpha^+, \omega) = -\frac{2}{j\omega} \sqrt{\frac{b}{\pi\sigma^2}} \frac{\phi_{-\beta\beta}(\alpha)}{\int_{-\alpha}^{\alpha} e^{b(y-x_0)^2/\sigma^2}}, \end{cases} \tag{5.64}$$

for the case of $\alpha < \beta$.

It is obvious that $G^0(-1, x, \omega) = G^0(x, \omega) - G^0(1, x, \omega)$ in both cases. The solutions of these problems coincide with the solution of problem (5.52) for edges E_2, E_3 and E_8, E_9. To completely solve problem (5.52), one should find the solution for the "central" edges E_{4-7}. However, it will be shown below that the cross-spectral density can be expressed in terms of the previously found functions, and consequently, the solution of problem (5.52) for these "central" edges is not necessary. Thus, the boundary-value problem (5.52) defined on the entire graph Z was reduced to the boundary-value problems defined on line intervals.

In the case of an OU input process, the specific form of the operator $\hat{L}_x$ is helpful in order to find explicit analytical solution to problem (5.52) in terms of parabolic cylinder functions [12]. Namely, one can observe that if a function $\tilde{f}$ satisfies the differential equation for the parabolic cylinder functions:

$$\frac{\partial^2 \tilde{f}}{\partial \tilde{x}^2}(\tilde{x}, \omega) + \left[-\frac{1}{4}\tilde{x}^2 + \left(\frac{1}{2} - j\frac{\omega}{b} \right) \right] \tilde{f}(\tilde{x}, \omega) = 0, \tag{5.65}$$

then $f(x, \omega) = \tilde{f}\left(\sqrt{2b(x - x_0)}/\sigma, \omega \right) e^{-b(x-x_0)^2/2\sigma^2}$ represents a solution to:

$$j\omega f(x, \omega) + \hat{L}_x f(x, \omega) = 0, \tag{5.66}$$

with $\hat{L}_x$ defined by Eq. (5.58). Let f_1 and f_2 be the solutions of Eq. (5.66) corresponding to the parabolic cylinder functions that vanish at $+\infty$ and $-\infty$, respectively. The solution of problem (5.52) on each graph edge can be expressed as a linear combination of these functions:

$$G^0((i_1, i_2, x), \omega) = \lambda_1(i_1, i_2, \omega)f_1(x, \omega) + \lambda_2(i_1, i_2, \omega)f_2(x, \omega) \qquad (5.67)$$

The coefficients $\lambda_1(i_1, i_2, \omega)$ and $\lambda_2(i_1, i_2, \omega)$ corresponding to each edge are found (for a given frequency) by matching the inhomogeneous "vertex" boundary conditions of the problem (5.52) (for that frequency). Thus, the analytical expression for the solution of the problem (5.52) can be expressed in terms of parabolic cylinder functions. Besides the importance in its own right, the described analytical approach can be used for the testing of the accuracy of numerical techniques.

Step 3

Using observation (3) from the previous section, the cross-spectral density $S_{\beta\alpha}(\omega)$ can be expressed as:

$$S_{\beta\alpha}(\omega) = -\frac{2}{\omega} \text{Im} \left\{ \int_{-\infty}^{\infty} \frac{\sigma^2}{2} \frac{\partial^2}{\partial x^2} \left(\sum_{i_\alpha, i_\beta} i_\beta G^0 \right) - b \frac{\partial}{\partial x} \left((x - x_0) \sum_{i_\alpha, i_\beta} i_\beta G^0 \right) dx \right\}$$

$$(5.68)$$

The derivatives in (5.68) can be integrated and appropriate vertex boundary conditions can be used for simplification. By using formulas (5.52) and (5.61–5.66), one can derive the following formula for the cross-spectral density, for $\alpha < \beta$:

$$S_{\beta\alpha}(\omega) = \frac{4\sigma\sqrt{b}}{\omega^2 \sqrt{\pi} \int_{-\beta}^{\beta} e^{b(y-x_0)^2/\sigma^2} dy}$$

$$- \frac{2\sigma^2}{\omega} \text{Im} \left[\frac{\partial G^0}{\partial x} (1, -\beta^+, \omega) - \frac{\partial G^0}{\partial x} (1, \beta^-, \omega) + \frac{\partial G^0}{\partial x} (\beta^+, \omega) \right], \qquad (5.69)$$

while for $\alpha > \beta$ we have:

$$S_{\beta\alpha}(\omega) = \frac{4\sigma\sqrt{b}}{\omega^2 \sqrt{\pi} \int_{-\alpha}^{\alpha} e^{b(y-x_0)^2/\sigma^2} dy}$$

$$- \frac{2\sigma^2}{\omega} \text{Im} \left\{ \sum_{i_\alpha} \left[\frac{\partial G^0}{\partial x} (1, i_\alpha, -\beta^+, \omega) - \frac{\partial G^0}{\partial x} (1, i_\alpha, \beta^-, \omega) \right] + \frac{\partial G^0}{\partial x} (\beta^+, \omega) \right\}.$$

$$(5.70)$$

According to Eq. (5.67), $G^0(\mathbf{y}, \omega)$ can be represented in terms of parabolic cylinder functions on each graph edge, hence explicit analytical formula in terms of parabolic cylinder functions can be given for cross-spectral density $S_{\beta\alpha}(\omega)$.

Results of the calculations for the cross-spectral density $S_{\beta\alpha}(\omega)$ using formulas (5.69) and (5.70) are presented in Fig. 5.10 where OU input process with $b = \sigma = 1$ and $x_0 = x_s = 0$ has been considered.

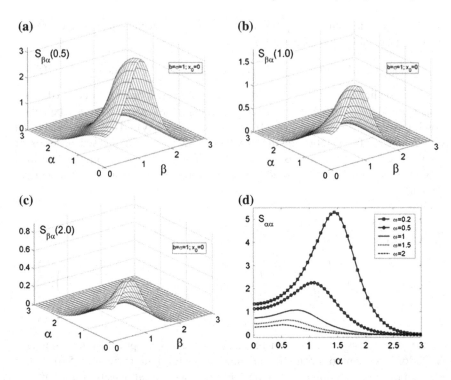

Fig. 5.10 Variation of cross spectral density $S_{\beta\alpha}$ with respects to the widths β and α of the two loops, for $\omega = 0.5$ (**a**), 1 (**b**), 2 (**c**); (**d**) Diagonal sections $S_{\alpha\alpha}$ are plotted for different frequencies f ($\omega = 2\pi f$). (© 2004 APS, [6])

Variations of the cross-spectral density $S_{\beta\alpha}(\omega)$ with respect to the widths β and α of the two loops are presented in Fig. 5.10a–c for selected values of the frequency while their diagonal sections ($S_{\alpha\alpha}(\omega)$) are compared in Fig. 5.10d for a better understanding of the relation between them. The cross-spectral density has negligible values outside of a finite region around the origin and this region becomes smaller when the frequency is increased. It can be clearly observed that the maximum of $S_{\alpha\alpha}(\omega)$ becomes more pronounced and it is shifted towards "wider loops" as the frequency is decreased. This suggests that two Preisach systems whose Preisach distributions coincide near the origin, should have approximately the same spectral noise densities for high frequencies. The computational results feature monotonic variations of $S_{\beta\alpha}(\omega)$ with respect to ω for fixed β and α, which leads to the conclusion that the spectral noise density $S_Y(\omega)$ of a Preisach system should be a decreasing function of frequency, regardless of the shape of the Preisach distribution. It is also expected that $S_{\beta\alpha}(\omega)$ is decreased for every β and α as x_0 is shifted from zero.

Step 4

Using formulas (5.69) and (5.70) for cross-spectral densities $S_{\beta\alpha}(\omega)$ in Eq. (5.35), the spectral density for the output process of a Preisach system

Fig. 5.11 Spectral density S_Y for a hysteretic system with uniform Preisach distribution for different values of input average value x_0 ($b = \sigma = 1$). (© 2004 APS, [6])

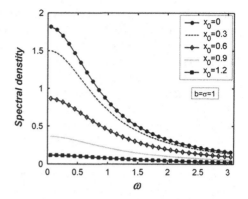

characterized by distribution μ and driven by an OU process can be found. Results of the calculations for the spectral density $S_Y(\omega)$ of a Preisach system with uniform distribution, $\mu(\alpha) = 1$, $\alpha \in (0, 1)$, are presented in Fig. 5.11.

As a final remark let us mention that in the limit of $b \to \infty$ with σ/b maintained constant, the Ornstein-Uhlenbeck process converges to a white noise process. Consequently, the output spectral density for the white noise input can be obtained either directly or as a limit of the result obtained for the Ornstein-Uhlenbeck process. In Refs. [18–20], Radons derived directly the output spectral density for a white noise input using a different technique and proved that long-time tails and even $1/f$ noise are quite general features of the class of symmetric Preisach models driven by uncorrelated noise.

5.3 Numerical Approach to Noise Spectral Analysis in Hysteretic Systems

For the numerical calculation of the output spectral density, a sufficiently large number of realizations of the noise input are generated according to the technique presented in Chap. 2 for the specific class of noise given in the problem. Then, the Fast-Fourier-Transform (FFT) technique is used to evaluate the spectral density of each output signal and average the output spectra to obtain their expected values. The power spectral density of the output signal is computed as:

$$S_Y(\omega) = \lim_{T \to \infty} \frac{E\left\{|Y_T(\omega)|^2\right\}}{2T} \qquad (5.71)$$

where $Y_T(\omega) = \int_{-T}^{T} y(t)e^{-j\omega t}dt$ is the "truncated" Fourier transform of the output signal $y(t)$. This approach has been implemented numerically in HysterSoft© and used to compute the noise spectral densities of the output signal for various hysteretic systems. For the simulations presented in this chapter, the spectrum of the output signal has been computed by averaging over 500 statistical (Monte-Carlo) simulations, which provided a very good accuracy of the results. The total time to evaluate $S_Y(\omega)$ on a one-processor computer operating at 3 GHz is less than a second for the energetic and Jiles-Atherton Models and less than a minute for the Preisach Model. The reliability of this numerical approach was successfully tested against several analytical results provided in the previous section (Sect. 5.1) for the hard limiter system, the hysteretic rectangular loop, and symmetric Preisach systems.

In this section, sample of the simulation results obtained using various hysteresis models driven by OU noise inputs are presented and analyzed. In the case of energetic, Jiles-Atherton, Preisach and Coleman-Hodgdon models, the parameters are chosen such that the corresponding major hysteretic loops have the same coercive input $x_c = 1.28$, output saturation $y_{sat} = 7.7 \times 10^5$, and output remanence $y_R = 4 \times 10^5$. These values, measured in A/m, characterize the major hysteretic loop of a permalloy ferrite [21]. The rest of models parameters and simulation are given in the subsections dedicated to a specific model. As discussed in the final subsections, the intrinsic differences between the algebraic, differential, and integral modeling of hysteresis are well exposed when the systems are driven by noisy inputs and their stochastic behaviors are compared against each other [22].

This analysis can be extended to the noise model of interest to the reader by selecting the noise model in HysterSoft© and running associated simulations.

5.3.1 Preisach Model

The Preisach distribution was identified on a discrete mesh of points using a set of first-order reversal-curves and employing Eqs. (1.17) and (1.18). This discrete distribution was then fitted to a 2-D normal distribution in order to speed up the computations:

$$P(\alpha, \beta) = \frac{y_{sat}S}{2\pi\, H_{\sigma i}H_{\sigma c}} \cdot \exp\left[-\frac{(\alpha + \beta - 2H_0)^2}{4H_{\sigma i}^2} - \frac{(\alpha - \beta)^2}{4H_{\sigma c}^2}\right], \qquad (5.72)$$

where $S = 0.88$, $H_{\sigma i} = 2.23$ A/m, $H_{\sigma c} = 0.49$ A/m, and the average value of the critical fields of the particles was found to be $H_0 = 1.9$ A/m. The reversible component of the Preisach distribution was also approximated by a normal distribution:

$$R(\alpha) = \frac{y_{sat}(1 - S)}{\sqrt{2\pi}\, H_{\sigma r}} \cdot \exp\left(-\frac{\alpha^2}{4H_{\sigma r}^2}\right), \qquad (5.73)$$

where $H_{\sigma r} = 2.12$. The initial hysteretic state in all simulations was assumed the zero-field anhysteretic curve (also known as the a.c. demagnetized state magnetism). In Fig. 5.12 the major hysteresis curve and an output realization of this Preisach system driven by an OU input are presented. The output spectra for different values of the noise strength are plotted in Fig. 5.13.

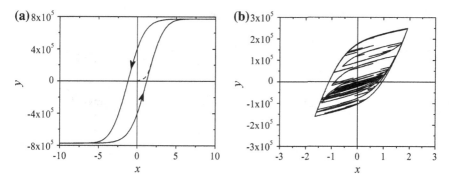

Fig. 5.12 Major hysteresis loop (**a**) and Minor hysteresis loops driven by an noisy input having an Ornstein-Uhlenbeck distribution with $b = \sigma = 1$ (**b**)

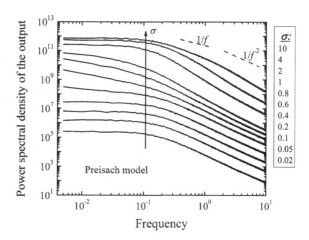

Fig. 5.13 Spectral density of the output of Preisach Model for different values of noise strength σ

5.3.2 Energetic Model

The parameters of the EM have been identified by using the technique presented in Sect. 1.4.3: $h = 0.4$, $k = 1.2$, $g = 8.24$, $c_r = 0.02$, $q = 10$, and $N_e = 3.5 \times 10^{-7}$. In Fig. 5.14 the major hysteresis curve and an output realization of this EM driven by an OU input are presented. The output spectra are plotted in Fig. 5.15 for different values of the noise strength.

5.3.3 Jiles-Atherton Model

The parameters of the JAM used in the simulations presented in this section are $k = 2.44$, $a = 4.36$, $\alpha = 1.7 \times 10^{-5}$, and $c = 0.49$ In Fig. 5.16 the major hysteresis

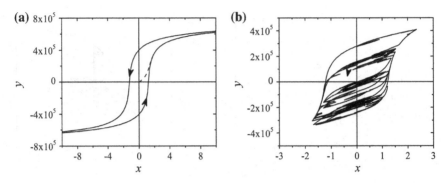

Fig. 5.14 Major hysteresis loop (**a**) and minor hysteresis loops driven by an noisy input having an Ornstein-Uhlenbeck distribution with $b = \sigma = 1$ (**b**)

Fig. 5.15 Spectral density of the output of energetic model for different values of noise strength σ

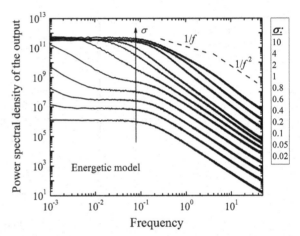

curve and an output realization of this JAM driven by an OU input are presented. The output spectra for different values of the noise strength are plotted in Fig. 5.17.

Figures 5.13, 5.15, and 5.17 show the results of the power spectral density of the magnetization computed by using the EM, JAM, and the PM, respectively, for different values of the diffusion coefficient (or noise strength) σ ranging from 40 for the top-most curves to 0.02 for the bottom-most curves. Let us remind that the spectrum of the magnetic field (noise input) has a Lorentzian-shape, which is flat in the low-frequency region and has $1/f^2$ decay for high-frequency region. As observed from these figures the last property is transferred by the hysteretic systems and the output spectra features a $1/f^2$ at high frequency region. It is interesting to note that all models predict a flat spectrum at low frequencies and large magnitudes of the input signal (large values of σ) and an increase of the low-frequency components for values of σ slightly lower than the coercive input, x_c. As opposed to the PM and EM, the JAM also predicts an increase in the power spectra

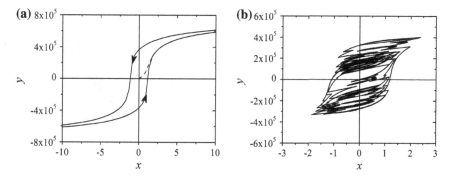

Fig. 5.16 Major hysteresis loop (**a**) and minor hysteresis loops driven by an noisy input having an Ornstein-Uhlenbeck distribution with $b = \sigma = 1$ (**b**)

Fig. 5.17 Spectral density of the output of Jiles-Atherton Model for different values of the noise strength σ

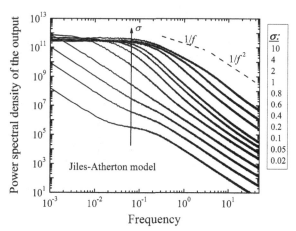

at low frequencies even for relatively low values of σ. In general, the hysteretic systems with OU input presented monotonic spectral densities and the increase in the input average x_o resulted in the decrease in the output noise.

5.3.4 Coleman-Hodgdon Model

The reverse Coleman-Hodgdon model with (1.117) with the material functions given by (1.127) and (1.128) is used below. The model parameters were identified as $A_1 = 1$, $A_2 = 2 \times 10^{-6}$, $A_3 = -0.7$, $A_4 = 0.01$, $\alpha = 8 \times 10^{-7}$, $x_{cl} = 2.5$, and $y_{cl} = 3 \times 10^6$. In Fig. 5.18 the major hysteresis curve and an output realization of this Coleman-Hodgdon system driven by an OU input are presented. The output spectra are plotted in Fig. 5.19 for different values of the noise strength.

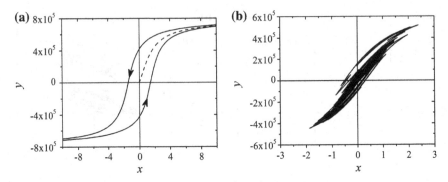

Fig. 5.18 Major hysteresis loop (**a**) and minor hysteresis loops driven by a noisy input having an Ornstein-Uhlenbeck distribution with $b = \sigma = 1$ (**b**)

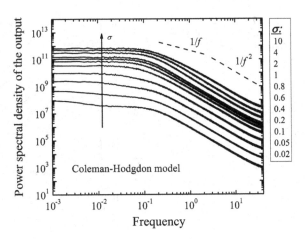

Fig. 5.19 Spectral density of the output of Coleman-Hodgdon model for different values of the noise strength σ

5.3.5 Bouc-Wen Model

Three sets of parameters have been used in the case of the Bouc-Wen model:

Set 1: $A = 2$, $\alpha = 0$, $\beta = 0.5$, $D = 2$, $\gamma = 0.1$, $k = 1$, $n = 1$;
Set 2: $A = 2$, $\alpha = 0$, $\beta = 0.5$, $D = 1$, $\gamma = 0.1$, $k = 1$, $n = 4$;
Set 3: $A = 2$, $\alpha = 0.7$, $\beta = 0.5$, $D = 1$, $\gamma = 0.1$, $k = 1$, $n = 1.1$.

These parameters correspond to hysteresis loops that resemble the stop operator often used in elasticity and plasticity (Figs. 5.20, 5.21, 5.22 and 5.23). Figures 5.20, 5.21, and 5.22 present the major hysteresis curves and the output realization for the Bouc-Wen model with the parameters given above. The output spectra are plotted in Fig. 5.15 for different values of the noise strength.

In conclusion, the numerical approach used here provides a relatively fast and reliable way to analyze the power spectral densities of complex hysteretic systems.

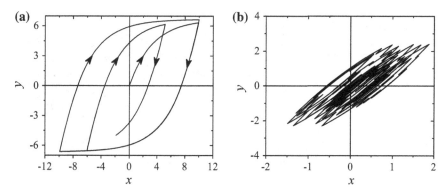

Fig. 5.20 Hysteresis loops driven by a deterministic input with simple monotonic variation (**a**) and a noisy input having an Ornstein-Uhlenbeck distribution with $b = \sigma = 1$ (**b**) the parameters of the Bouc-Wen model are defined in Set 1

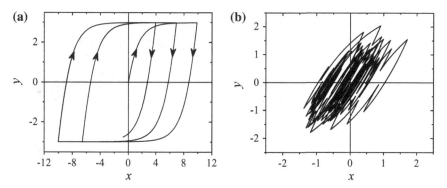

Fig. 5.21 Hysteresis loops driven by a deterministic input with simple monotonic variation (**a**) and a noisy input having an Ornstein-Uhlenbeck distribution with $b = \sigma = 1$ (**b**) the parameters of the Bouc-Wen model are defined in Set 2

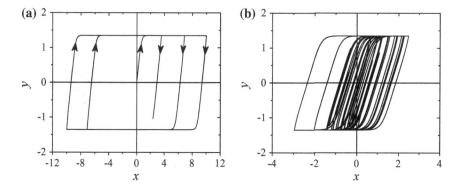

Fig. 5.22 Hysteresis loops driven by a deterministic input with simple monotonic variation (**a**) and a noisy input having an Ornstein-Uhlenbeck distribution with $b = \sigma = 1$ (**b**) the parameters of the Bouc-Wen model are defined in Set 3

Fig. 5.23 Spectral densities of the output of Bouc-Wen model for different values of the noise strength σ in the case of the model parameters defined by Set 1 (**a**), Set 2 (**b**), and Set 3 (**c**)

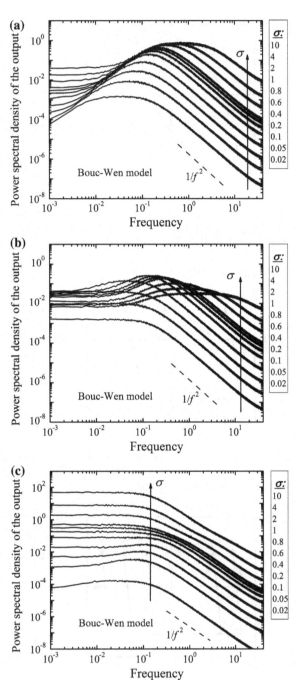

According to our analysis, the output spectra deviate significantly from the Lorentzian shape of the input process for values of the diffusion coefficient near and smaller than the coercive field. The intrinsic differences between the transcendental, differential, and integral modeling of hysteresis yield significantly different spectra at low frequency region, which reflect the diverse long-time correlation behavior. It is also apparent from this study that the spectral analysis is a powerful characterization tool that can be used to design filters based on hysteretic systems.

References

1. Gardiner, C. (2009). *Stochastic methods: A handbook for the natural and social sciences*. Berlin: Springer.
2. Freidlin, M. I., & Wentzell, A. D. (1993). Diffusion processes on graphs and the averaging principle. *Annals of Probability, 21*(4), 2215–2245.
3. Freidlin, M. I. (1996). *Markov processes and differential equations: Asymptotic problems*. Berlin: Springer.
4. Freidlin, M. I., Mayergoyz, I. D., & Pfeiffer, R. (2000). Noise in hysteretic systems and stochastic processes on graphs. *Physical Review E, 62*, 1850–1856.
5. Mayergoyz, I., & Dimian, M. (2003). Analysis of spectral noise density of hysteretic systems driven by stochastic processes. *Journal of Applied Physics, 93*(10), 6826–6828.
6. Dimian, M., & Mayergoyz, I. D. (2004). Spectral density analysis of nonlinear hysteretic systems. *Physical Review E, 70*, Article 046124.
7. Dimian, M. (2008). Extracting energy from noise: noise benefits in hysteretic systems. *NANO: Brief reviews and reports, 3*(5), 391–397.
8. Dimian, M., Gîndulescu, A., & Andrei, P. (2010). Influence of noise temporal correlation on magnetization spectra and thermal relaxations in soft magnetic materials. *IEEE Transactions on Magnetics, 46*(2), 266–269.
9. Mayergoyz, I. D. (2003). *Mathematical models of hysteresis and their applications* (2nd ed.). New York: Academic Press.
10. Melnikov, V. I. (1993). Schmitt trigger: A solvable model of stochastic resonance. *Physical Review E, 48*(4), 2481–2489.
11. Lindner, B., Garcia-Ojalvo, J., Neiman, A., & Schimansky-Geier, L. (2004). Effects of noise in excitable systems. *Physics Reports, 392*, 321–424.
12. Abramowitz, M., & Stegun, I. (Eds.). (1972). *Handbook of mathematical functions*. New York: Dover Publications.
13. Papoulis, A. (2002). *Probability, random variables and stochastic processes*. New York: McGraw-Hill.
14. van Vleck, J. H. (1943). The spectrum of clipped noise. *RRL Report* 51.
15. Pikovsky, A. S., & Kurths, J. (1997). Coherence resonance in a noise-driven excitable system. *Physical Review Letters, 78*, 775–778.
16. Korman, C. E., & Mayergoyz, I. D. (1996). Semiconductor noise in the framework of semiclassical transport. *Physical Review B, 54*, 17620–17627.
17. Berezin, I. S., & Zhidkov, N. P. (1965). *Computing methods*. London: Pergamon.
18. Radons, G. (2008). Hysteresis-induced long-time tails. *Physical Review Letters, 100*, Article 240602.
19. Radons, G. (2008). Spectral properties of the Preisach hysteresis model with random input. I. General results. *Physical Review E, 77*, Article 061133.
20. Radons, G. (2008). Spectral properties of the Preisach hysteresis model with random input. II. Universality classes for symmetric elementary loops. *Physical Review E, 77*, Article 061134.

21. Andrei, P., & Adedoyin, A. (2009). Noniterative parameter identification technique for the energetic model of hysteresis. *Journal of Applied Physics, 105*, Article 07D523.
22. Adedoyin, A., Dimian, M., & Andrei, P. (2009). Analysis of noise spectral density for phenomenological models of hysteresis. *IEEE Transactions on Magnetics, 45*(10), 3934–3937.

Chapter 6
Constructive Effects of Noise in Hysteretic Systems

6.1 Introduction

Since everyone senses the negative effects of noise, its potential benefits seem counterintuitive. As it was discussed in Sect. 2.1.1, intentionally applied noise is helping sub-threshold constant signals to surpass the activation threshold of a system. Thus, the system can switch from one metastable state (let us say 0-state) to the other (1-state) with the help of noise, even when the constant signal acting on this switching is not strong enough to produce it alone. Noise is also acting in other instants of time towards diminishing the constant signal but that does not have an effect on the system state. In conclusion, thermal noise may play a positive role in achieving higher storage density in magnetic recording nanotechnology by using heat assisted magnetic recording (HAMR). In 2012, Seagate Technology has demonstrated an operating prototype of a hard disk drive based on HAMR featuring 1 Tb/in^2 areal storage density, which is significantly higher than 600 Gb/in^2 used in today's hard disk drives.

What is more difficult to grasp regarding the benefits of noise is the help provided by the random nature of noise. Let us start with an example which is not particularly included in the area of hysteresis but might open some perspective in our intuitive approach to these phenomena. Everybody is familiar to hand sieving for separating particles of different sizes, so it is apparent that a purely noisy shake of the sieve will have constructive effects in practice, such as sifting flour or separating stones from sand. The first known example of using noise to enhance the performance of modern technologies is given in the 1940's by British naval air fleet, when they realized that their navigations systems perform better when the airplanes were flying than when they were on the ground. It was determined that the vibrations from plane engine help the rigid components (cranks, gears, cogs, etc.) of the navigation system to move smoother avoiding, for example, the possibility for some of them to stick together. As a result, small motors were installed on all British navigation systems just to provide the vibrations needed by the rigid mechanisms to operate more fluidly [1]. This concept was later generalized to signal processing becoming a fundamental technique, known as dithering, that is

M. Dimian and P. Andrei, *Noise-Driven Phenomena in Hysteretic Systems,*
Signals and Communication Technology 218, DOI: 10.1007/978-1-4614-1374-5_6,
© Springer Science+Business Media New York 2014

today used in numerous digital systems from various areas such as audio and video processing, communications, radio-location and detection, seismology.

Another major direction in the analysis of noise benefits is provided by stochastic resonance phenomena. Let us consider a bistable system and a sinusoidal input that is not strong enough to switch the system from one metastable state to another. By applying a relatively small amount of noise, transitions between the states may happen but are rare events, while by applying a relatively large amount of noise, the transitions would be too frequent and irregular, without relevant correlation to the initial sinusoidal input. However, for an intermediate level of noise, some regularity of the transitions can be observed in correlation to the initial sinusoidal frequency. This noise-induced phenomenon is coined as stochastic resonance. The idea was proposed almost 30 years ago in a model for climate change and it spread very fast across various areas of science and engineering.

6.2 Dithering

In order to illustrate the dithering technique, let us consider an analog-to-digital convertor (ADC). Although the information is mostly generated in an analog form, its transmission is desired in a digital form due to better performances in encoding, compressing, and encrypting the digital signal. Thus, the digital signal is better protected against communication channel noise and fading, needs a lower transmission rate, and can better prevent eavesdropping and interception. The ADC involves sampling (time digitization) and quantization (amplitude digitization) of the analog signal. The quantization errors are inherent and the dithering technique is trying to decrease this error with the help of noise. For example, let us consider two quantization levels (0 and 1) and the quantization rule of truncating to zero all positive sub-unitary values smaller 0.5 and rounding off to one all positive sub-unitary values larger or equal to 0.5. By applying noise, an initial value which is below 0.5 threshold may get above this threshold and be rounded off to 1. Thus, when the sampling rate is high compared to the rate of analog signal change, consecutive values of the initial signal may be approximated by 0 or 1 even when they are all below 0.5. This provides a local average of the digital dithered signal closer to the initial local average of the analog signal than the digitized non-dithered signal (which is zero for this interval of time). When the digital signal is received and smoothed out, these local averages play an important role in the smoothing process, so the resulting signal coming from the digitized dithered signal may be a better approximation of the initial analog signal than the digitized non-dithered signal. In conclusion, the quantization error may be reduced by this random selection of 0 and 1 with a probability depending on the difference between the given signal value and the 0.5 threshold.

The spectral comparison between the digitized forms of the initial and dithered signal provides a more compelling support for the previous conclusion as it is discussed in the next example. However, we invite the reader to consult references [1–3]

for a more rigorous and detailed analysis of this idea. A digitized form of a pure
sinusoidal signal is presented in Fig. 6.1a while its power spectrum is presented in
Fig. 6.1c. In addition to the peak corresponding to the original sinusoidal signal
(located at 1 MHz) multiple peaks with significant height can be identified in this
spectrum. They correspond to the higher order harmonics generated by the digitization process. When the dither noise is added to the original signal, the digitization
process of this noisy sinusoidal signal generates a digitized form presented in
Fig. 6.1b and its corresponding power spectrum presented in Fig. 6.1d. By comparing the two power spectra, it is apparent that no major peaks can be identified in
the dithered spectrum except the one corresponding to the original sinusoidal signal.
However, the ground noise level has been increased in the dithered signal. In conclusion, the dithered technique is trying to reduce the higher order harmonics generated by the digitization process on the expense of increasing the level of ground
noise.

The dithering technique has been extensively used in ADC to decrease the
distortion of low-frequency signal due to digitization process and its applications
have been mainly focused on audio signals. During the last years the technique
was successfully generalized to higher frequencies and the area of applications was

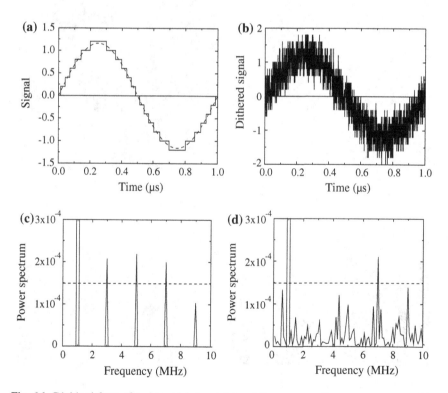

Fig. 6.1 Digitized form of a sinusoidal signal (**a**) and the corresponding power spectrum (**c**);
Digitized form of the same sinusoidal signal with dither noise (**b**) and the corresponding spectrum (**d**)

significantly extended [4–6]. As an example of this extended range of applications, let us consider dithering in digital graphics. A standard image used to test processing algorithm is Lena, which is presented in Fig. 6.2a. It is a grey version of this image with grey level varying from 0 (white) to 1 (black). A standard purely black and white version, where all pixels with grey intensity lower than one half are changed to white and all others to black, is presented in Fig. 6.2b. Now, let us consider dithered versions of the grey picture where zero-average Gaussian noise is added to the pixels. The standard procedure to generate a pure black-and-white image is next applied. Two realizations of the dithered image with Gaussian noise having different strength (standard deviations) are presented in Fig. 6.2c and 6.2d. As a general rule reflected by this example, zero or low noise generates a blotchy black-and-white image since large areas are represented by a single color; a large amount of noise added to the original image generates a grainy black-and-white image lacking significant details of the original; nevertheless, an intermediate amount of noise generates a more accurate representation of the original than the non-dithered version. This dithering technique can be obviously applied to colored

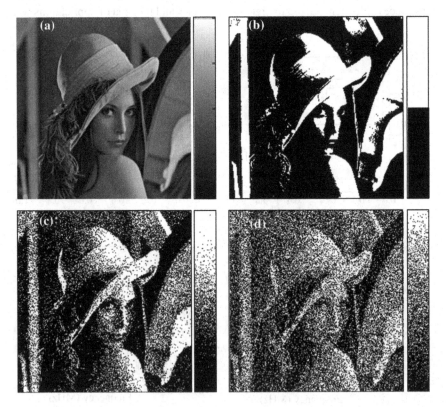

Fig. 6.2 (a) Grey version of Lena standard image; (b) Black-and-white version of the original image; (c) Black-and-white version of a dithered image with moderate noise; (d) Black-and-white version of a dithered image with large noise. (based on Ref. [7])

pictures when several quantized colors are used to generate an approximate version of the original image. By taking advantage of the human eye tendency to mix two colors in close proximity to one another, the noise-added image looks grainy but also more accurate than non-dithered version since the colors blend together more smoothly. A detailed account of dithering in digital graphics can be found in Ref. [6].

6.3 Noise Induced Resonances in Bistable Systems

In Sects. 4.2.2 and 5.1.4, noise induced relaxations and spectral noise densities for a bistable hysteretic system driven solely by noise have been computed and analyzed. Here, the discussion is extended for bistable hysteretic systems driven by a noisy sinusoidal signal:

$$x(t) = X_0 \sin(2\pi f\, t) + n(t), \tag{6.1}$$

where $n(t)$ is a noise component superimposed on a sinusoidal component of frequency f and magnitude X_0. The case of symmetric hysteretic loop is considered for the sake of clarity but the analysis can be extended to non-symmetric cases. Let us assume that the magnitude X_0 is smaller than the hysteretic threshold, so the deterministic signal is not strong enough to generate any switching in the system state. With the help of a weak noise, the switching is possible but as a rare event and at irregular intervals of time. This case is illustrated in Fig. 6.3. When strong noise is added to the sinusoidal input, the switching in the system state would be too frequent and irregular, with little correlation to the deterministic input. This case is illustrated in Fig. 6.4. However, for an intermediate level of noise, some regularity of the transitions can be observed in correlation to the initial sinusoidal frequency, as illustrated by Fig. 6.5. This noise-induced regular transitions in the presence of an oscillatory signal is known as stochastic resonance.

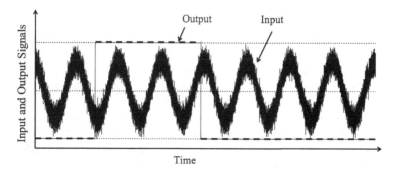

Fig. 6.3 A sub-threshold sinusoidal input with weak noise and the resulting output of the bistable hysteretic system

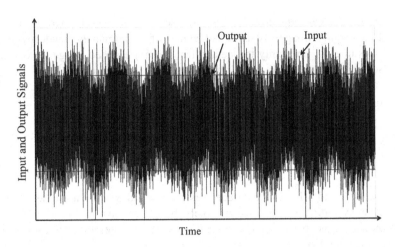

Fig. 6.4 A sub-threshold sinusoidal input with strong noise and the resulting output of the bistable hysteretic system

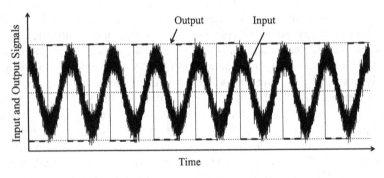

Fig. 6.5 A sub-threshold sinusoidal input with an intermediate level of noise and the resulting output of the bistable hysteretic system

This stochastic resonance concept was first proposed in [8] by Benzi and his collaborators for the Brownian motion in a Landau potential (see 2.1.6) which is also subject to a small periodic forcing. They proved that a peak in the power spectrum appears which is absent when either the forcing term or noise is absent. It was then applied to a stochastically perturbed Budyko-Sellers model for the variation of the global earth temperature in an attempt to explain the almost periodic occurrence of ice age [9]. Although subsequent experimental data did not confirm this hypothesis regarding climate change, the concept of stochastic resonance has spread fast in many other areas of science and engineering. The experimental confirmations coming few years later in a study of a bistable electronic system [10] and a bidirectional ring laser [11] have provided the needed confidence to the extensive theoretical analysis. Another significant breakthrough in this area of research came with proof of stochastic resonance in biological systems such as crayfish mechanoreceptors [12] and cricket cercal sensory system [13]. A detailed account of the developments in this area for the first two decades can be found in the review paper [14] and the book of Ando and Graziani [15]. The latter also provides a Matlab-based simulation software for the analysis of stochastic resonance in bitable and quasi-linear systems.

The last decade has been marked by numerous studies on the constructive role of noise in nervous systems, ranging from synapsis to cortex, as well as by extensive theoretical search on finding new systems manifesting stochastic resonance along with the general conditions under which such phenomena occur [16–21].

A significant element has been added to this field by proving that several dynamical systems can exhibit a coherent behavior solely induced by noise, phenomenon that was first named *stochastic resonance without external periodic driving* [22], and later *autonomous stochastic resonance* [23] and *coherence resonance* [24, 25]. The last term has been adopted by many other authors during the last years so it is also used in this book to describe these noise induced phenomena. Coherence resonance was demonstrated in a number of experimental studies including electronic circuits [26], laser systems [27], chemical reaction systems [28, 29], and neural systems [30]. In Sect. 5.1.4, it has been shown that coherence resonance may take place in bistable hysteretic systems with the internal noise characteristics depending on the system state.

From a theoretical point of view, most of these studies can be framed into two-state models or simple variants thereof, while complex multi-stable systems are rarely addressed. The extensive experience of hysteretic modeling community [31] may significantly contribute to the analysis of noise benefits in systems with complex metastable configurations. So, one of the goals of this book is to provide a unitary framework for studying noise-induced phenomena in complex hysteretic systems and its implementation in an open-access academic software. By using this framework and HysterSoft©, various differential, integral, and algebraic models of hysteresis in an arbitrary colored noise environment can be considered by graduate students, as well as advanced researchers, for the analysis of constructive effects of noise in hysteretic system with complex metastable state configurations. Several examples are provided in the next section following the line of the recent articles [32–34] published by our group.

6.4 Noise-Induced Resonance in Complex Hysteretic Systems

In the first part of this section, the noise component of input, see (6.1), is a white Gaussian noise which is simulated by a discrete-time i.i.d. random process, normally distributed with zero average and standard deviation equal to σ.

First class of systems addressed in this analysis is defined by the Preisach formalism as a weighted superposition of elementary hysteron operators described by rectangular loops. An extensive account of Preisach model (PM) and its variants can be found in Sect. 1.2. Here we consider the classical PM with normal distributions along the interaction and critical fields with parameters $\sigma_\xi = 0.1$, $\sigma_\eta = 0.1$, $\eta_0 = \sqrt{2}$, $y_{sat} = 1$, and $S = 1$, where $S = y_{si}/y_{sat}$ is the squareness factor. Examples of the input–output diagrams simulated by using PM driven by a sinusoidal input with and without noise are presented in Fig. 6.6 (left). The topic of magnetic stochastic resonance in PM was addressed for the first time in [35, 36].

The second hysteresis model considered in this analysis is the Jiles-Atherton model (JAM), which has the input–output relation described by a differential equation which involves the monotonicity of the input variables. A detailed description of this model can be found in Sect. 1.3. Examples of the input–output diagrams simulated by using the JAM driven by a sinusoidal input with and without noise are presented in Fig. 6.6 (right) for the following model parameters: $k = 1.3$, $c = 0.08$, $\alpha = 0$, $a = 0.5$, and $y_{sat} = 1$.

The third class of systems are using the energetic model (EM), in which the output is related to the input by a transcendental algebraic equation involving the past values of reversal points of the output, as presented in Sect. 1.4. Examples of the input–output diagrams simulated by using EM driven by a sinusoidal input with and without noise are presented in Fig. 6.7 (left), where we have used the following model parameters: $c_r = 0.1$, $N_e = 0$, $q = 11$, $h = 0.4$, and $g = 8.24$; the other parameters have been selected in order to achieve a coercive field of 1 and $y_{sat} = 1$.

The last class considered in this analysis is described by Coleman-Hodgdon Model (CHM) in which the input–output relation is a differential one, as described in Sect. 1.6. In our simulations we use, $A_1 = 0.2$, $A_2 = 1$, $A_3 = -2$, $A_4 = 0.25$, $\alpha = 1$, $\mu_s = 1$, and $y_{bp} = y_{cl} = 2.5$. Examples of the input–output diagrams simulated by using CHM driven by a sinusoidal input with and without noise are presented in Fig. 6.7 (right).

Let us note that significant amplifications of the output signal can be observed in Figs. 6.6 and 6.7 for all four types of hysteretic models in the presence of noise.

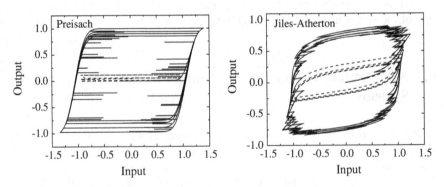

Fig. 6.6 Input-output diagrams for Preisach (*left*) and Jiles-Atherton (*right*) models driven by a sinusoidal input without (*dotted lines*) and with noise (*continuous lines*)

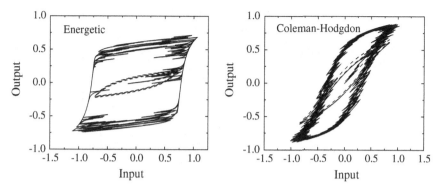

Fig. 6.7 Input-output diagrams for Energetic (*left*) and Coleman-Hodgdon (*right*) models driven by a sinusoidal input without (*dotted lines*) and with noise (*continuous lines*)

High amplification and regularity of the output have been identified when the coercive field of the major loop, the magnitude of the deterministic signal, and the noise strength satisfy certain conditions. In order to find these conditions for the stochastic resonance behavior, the output signal amplification (SA) and signal-to-noise ratio (SNR) have been computed. An intermediate step in this attempt was to compute the output spectral density. To this purpose, a sufficiently large number of realizations for the stochastic input underlying (6.1) have been generated and the corresponding output signals have been computed by using the hysteretic model under consideration. Then, the Fast Fourier Transform technique was applied to each output signal and the results were averaged out over the all realizations. Additional details regarding the calculation for output spectral density for complex hysteretic systems have been presented in the previous Chapter.

Finally, the SNR and the SA were computed by evaluating the Fourier components of the output corresponding to frequency f of the input sinusoid. The SNR was computed as the ratio between the mean μ_1 and the standard deviation σ_1 of the averaged signal:

$$SNR = 10\log(\mu_1/\sigma_1), \tag{6.2}$$

while the SA as the ratio between mean μ_1 and the magnitude of the Fourier component of the output signal corresponding to the same frequency f but computed in the absence of noise:

$$SA = 10\log(\mu_1/\mu_{10}), \tag{6.3}$$

This noise characterization technique is implemented in HysterSoft© and can be used by the reader for the systems of interest. The sample models used in this comparison have major loops with both the critical field and saturation equal to 1, as shown in Fig. 6.8. The Coleman-Hodgdon and Energetic examples have been chosen to have similar major loops. However, the minor loops are relatively different from one class to another due to the fundamental differences between

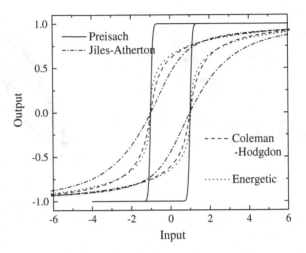

Fig. 6.8 The major hysteretic loops for four models used in our comparison

these models, and lead to significantly different characteristics of noise induced amplification, as it is apparent by comparing the simulations shown in Fig. 6.9. The signal amplification decreases with the increase in sinusoid amplitude, as expected, but the resonant noise strength increases (Coleman-Hodgdon) or stays constant (Energetic). Let us remark that the variation of output signal amplification with noise strength features a maximum even for sinusoid amplitude higher than the critical field of the hysteretic model (Fig. 6.10).

Due to the high degree of hysteretic loop squareness for the selected PM, the signal amplification is much higher than the SA of the other models used in our comparison. However, the input noise is also significantly amplified by this PM and consequently, the range of SNR is similar to all models, as it is also apparent from Fig. 6.11. The high degree of loop squareness makes this system behavior

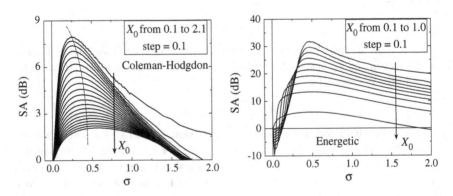

Fig. 6.9 Variation of output signal amplification with respect to noise strength σ for different values of sinusoid amplitude X_0 in the case of selected Coleman-Hodgdon and Energetics models

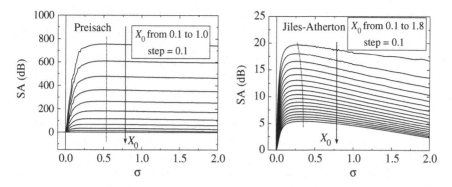

Fig. 6.10 Variation of output signal amplification with respect to noise strength σ for different values of sinusoid amplitude X_0 in the case of selected Preisach and Jiles-Atherton models

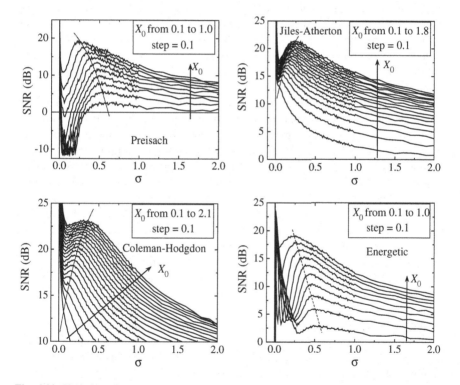

Fig. 6.11 Variation of output signal-to-noise ratio with respect to noise strength σ for selected values of sinusoid amplitude X_0 in the case of selected Preisach, Jiles-Atherton, Coleman-Hodgdon and Energetics models

rather similar to the behavior of bistable systems, which have been extensively analyzed over the last years. Thus, this PM shape of SNR variation with the noise strength σ and the resonant value of σ for weak periodic signals can be approximated by the analytical results based on small signal analysis of bistable systems and the exit problem approach applied to a Brownian motion in a semi-finite interval or a potential well [37–39].

SNR characteristics for the models considered in this analysis indicate a clear maximum at non-zero noise strength, which is a fingerprint of stochastic resonance, but the resonant noise strength variation with the deterministic input amplitude follows significantly different monotonic paths for each type of models. Once the degree of loop squareness is reduced and the complexity of metastable state landscape of a general hysteretic system is manifested, the noise induced resonance characteristics have significantly different features than the ones known from usual stochastic resonance phenomena obtained for bistable systems and their variants.

It is also clear from this analysis that systems having approximately the same major hysteretic loops (as it can be seen Fig. 6.9 for Energetic and Coleman-Hodgdon examples) respond in a completely different manner to a noisy input. By exploring higher order reversals curves and minor hysteretic loops, a noisy input emphasizes the fundamental differences between the hysteresis models related to memory, accommodation, thermal relaxation, and other relevant characteristics.

Important changes have also been observed in the input–output diagrams and signal amplification characteristics when applying the *dynamic Jiles-Atherton model* to investigate stochastic resonance phenomena. In Fig. 6.11 are presented four examples of input–output diagrams obtained for the dynamic Jiles-Atherton model driven by sinusoidal inputs with noise, when the relaxation time parameter τ is considered 0, 0.2, 1, and 2, respectively. The sinusoidal amplitude X_0 in these examples is taken 1.2, so it is larger than the critical field of the model and yet important signal amplification is happening. The variation of output signal amplification and signal-to-noise-ratio with respect to input noise strength for selected sub-threshold values of sinusoid amplitude is represented in Fig. 6.12 in the case of τ equal to 0 and 0.1. Although SA changes with the change of relaxation rate τ, the SNR remains almost constant, the observation confirm by simulations made for higher values of τ, as well. Consequently, these dynamic Jiles-Atherton models maintain the SNR features of static Jiles-Atherton model.

In conclusion, white noise can have a constructive effect in various complex hysteretic systems, activating some kind of resonance response. The quantities used to characterize this behavior are signal amplification and signal-to-noise ratio, which displays a maximum at the resonance noise strength. The statistical technique includes various algebraic, differential and integral models of hysteresis and is implemented in HysterSoft© (Fig. 6.13).

In the final part of this Chapter, the influence of noise color on noise-induced resonance phenomena in complex hysteretic systems is investigated. In general, colored noise is the complementary notion of white noise including noises with flat spectrum only on a finite frequency band and noises with non-flat spectrum. Practically all real noises are colored to some degree, but white noise is much more convenient for theoretical analysis due to its mathematical simplicity. In Chap. 2, the main types of colored noise are discussed along with the numerical

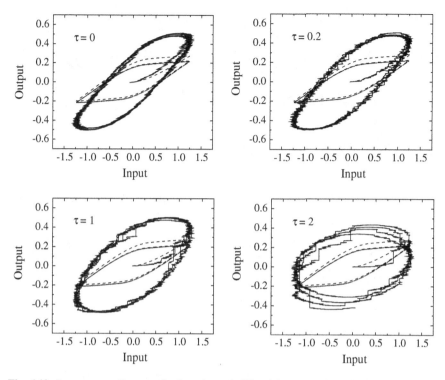

Fig. 6.12 Input-output diagrams for four dynamic Jiles-Atherton models driven by sinusoidal inputs with noise. The relaxation time τ is considered 0, 0.2, 1, and 2, respectively

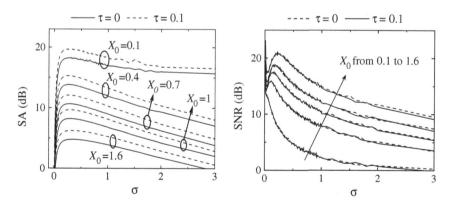

Fig. 6.13 Variation of output signal amplification (*left*) and signal-to-noise-ratio (*right*) with respect to input noise strength σ for selected values of sinusoid amplitude X_0 in the case of dynamic Jiles-Atherton for two selected values of relaxation time τ

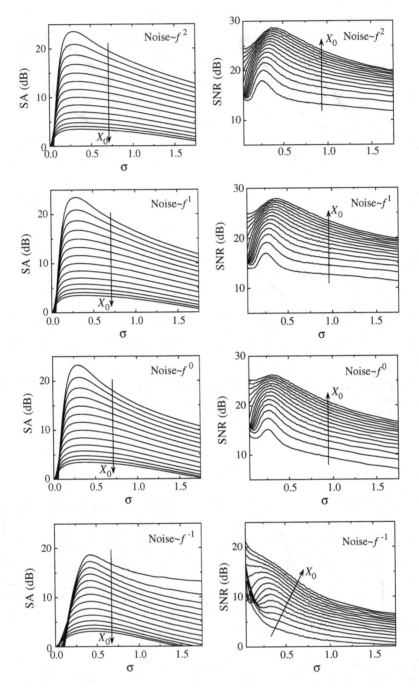

Fig. 6.14 Variation of output SA and SNR with respect to noise strength σ for selected values of deterministic input amplitude X_0 in the case of EM subject to *violet* (f^2), *white* (f^0), *pink* (f^{-1}), and *red* (f^{-2}) noise

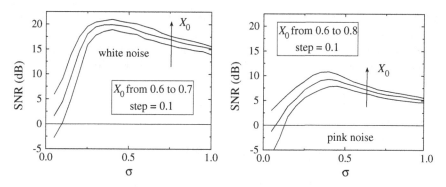

Fig. 6.15 Variation of output SNR with respect to noise strength σ for selected values of deterministic input amplitude X_0 in the case of a PM subject to *white* (f^0) and *pink* (f^{-1}) noise

techniques used to simulate them. Here, the colored noise is numerically generated according to the method described in Sect. 2.1.8 starting from a discrete-time i.i.d. random process, normally distributed with zero average and standard deviation denoted by σ. For a consistent comparison between the effects of different types of colored noise, the noise power is normalized to unity in all cases. The noise spectrum is considered proportional to f^α but arbitrary colored noise can be by considered and studied by the reader using HysterSoft©.

In the case of a bistable hysteretic systems, the computational results for SNR characteristics indicates a maximum at non-zero noise strength for sub-threshold sinusoidal signal, which is a fingerprint of stochastic resonance [40, 41]. The resonant noise strength is decreasing with the increase in the amplitude of external oscillation but this variation follows quantitatively different paths for each case of noise color analyzed. For a fixed external amplitude, the maximum of the SNR decreases and moves towards large noise intensities when decreasing the power coefficient α [34]. As expected, the constructive effects of noise disappear in bistable hysteretic systems when external signal is strong enough to overcome the system threshold (supra-threshold sinusoidal signals). Similar conclusions were obtained by using different techniques in References [11, 12], where the case $\alpha \leq 0$ was analyzed. However, the characteristics of this behavior change significantly when multi-stable hysteretic systems are analyzed.

Samples of the simulation results for the EM driven by noisy oscillatory input are presented in Fig. 6.14, where SA and SNR are plotted against the noise strength for selected values of sinusoid amplitude in the case of blue, white, pink, and red noise. Samples of SNR characteristics for the PM are plotted in Fig. 6.15 in the case of white and pink noise. The models have been selected to describe a hysteretic system in which the major loop has both the critical field and saturation equal to 1, as it was in the rectangular loop case. However, smoother transitions from negative to positive saturations make these complex systems exhibit SR for sub-critical as well as for supra-critical field values of sinusoid amplitude. While PM maintains the rectangular loop feature of decreasing the resonant noise strength when the amplitude of external oscillation is increased, the EM has an opposite behavior for positive values of power coefficient α. As it is also apparent from the last two figures, for a fixed external amplitude, the maximum of the SNR decreases and moves toward large noise intensities when decreasing α.

References

1. Aldrich, N. (2004). *Digital audio explained: for the audio engineer* (2nd ed.). Fort Wayne: Sweetwater Sound Inc.
2. Lyons, R. G. (ed.) (2012). *Streamlining digital signal processing: A tricks of the trade guidebook* (2nd ed.). Hoboken: John Wiley &Sons.
3. Bohn, D. (2006). *Pro audio reference* (2nd ed.). Mukilteo: Rane Corp.
4. Kosko, B. (2006). *Noise*. New York: Viking/Penguin.
5. Malekzadeh, F. A., Mahmoudi, R., van Roermund, A. H. M. (2012). *Analog dithering techniques for wireless transmitters*. New York: Springer.
6. Velho, L., Frery, A. C., Gomes, J. (2009). *Image processing for computer graphics and vision* (2nd ed.). London: Springer.
7. Hocevar, S. (2013). *The science behind colour ASCII art*. Libcaca software library.
8. Benzi, R., Sutera, A., & Vulpiani, A. (1981). The mechanism of stochastic resonance. *Journal of Physics A: Mathematical and General, 14*, L453–L457.
9. Benzi, R., Parisi, G., Sutera, A., & Vulpiani, A. (1982). Stochastic resonance in climatic change. *Tellus, 34*, 10–16.
10. Fauve, S., & Heslot, F. (1983). Stochastic resonance in a bistable system. *Physical Letters A, 97*, 5–7.
11. McNamara, B., Wiesenfeld, K., & Roy, R. (1988). Observation of stochastic resonance in a ring laser. *Physical Review Letters, 60*, 2626–2629.
12. Douglass, J. K., Wilkens, L., Pantazelou, E., & Moss, F. (1993). Noise enhancement of information transfer in crayfish mechanoreceptors by stochastic resonance. *Nature, 365*, 337–340.
13. Levin, J. E., & Miller, J. P. (1996). Broadband neural encoding in the cricket cercal sensory system enhanced by Stochastic Resonance. *Nature, 380*, 165–168.
14. Gammaitoni, L., Hanggi, P., Jung, P., & Marchesoni, F. (1998). Stochastic resonance. *Reviews of Modern Physics, 70*, 223–287.
15. Ando, B., & Graziani, S. (2000). *Stochastic resonance*. Boston: Kluwer Academic Publishers.
16. Lindner, B., Garcia-Ojalvo, J., Neiman, A., Schimansky-Geier, L. (2004). Effects of noise in excitable systems. *Physics Reports, 392*, 321–424.
17. Sagues, F., Sancho, J. M., & Garcia-Ojalvo, J. (2007). Spatiotemporal order out of noise. *Reviews of Modern Physics, 79*, 829–882.
18. Gammaitoni, L., Hänggi, P., Jung, P., & Marchesoni, F. (2009). Stochastic resonance: A remarkable idea that changed our perception of noise. *European Physical Journal B, 69*, 1–3.
19. Linder, B., Erdmann, U., Sokolov I. M. (Eds.). (2010). The dynamics of nonlinear stochastic systems. *The European Physical Journal, Special Topics, 187*, 1–274.
20. Berglund, N., Gentz, B. (2010). *Noise-induced phenomena in slow-fast dynamical systems: a sample-paths approach*. London: Springer.
21. McDonnell, M. D., Stocks, N. G., Pearce, C. E. M., Abbott, D. (2008). *Stochastic resonance: From suprathreshold stochastic resonance to stochastic signal quantization*. New York: Cambridge University Press.
22. Gang, H., Haken, H., & Fagen, X. (1993). Stochastic resonance without external periodic force. *Physical Review Letters, 77*, 1925–1928.

23. Longtin, A. (1997). Autonomous stochastic resonance in bursting neurons. *Physical Review E, 55*, 868–876.

24. Pikovsky, A. S., & Kurths, J. (1997). Coherence resonance in a noise-driven excitable system. *Physical Review Letters, 78*, 775–778.

25. Neiman, A., Saparin, P. I., & Stone, L. (1997). Coherence resonance at noisy precursors of bifurcations in nonlinear dynamical systems. *Physical Review E, 56*, 270.

26. Postnov, D. E., Han, S. K., Yim, T. G., & Sosnovtseva, O. V. (1999). Experimental observation of coherence resonance in cascaded excitable systems. *Physical Review E, 59*, R3791–R3794.

27. Giacomelli, G., Giudici, M., Balle, S., & Tredicce, J. R. (2000). Experimental evidence of coherence resonance in an optical system. *Physical Review Letters, 84*, 3298–3301.

28. Miyakawa, K., Isikawa, H. (2002). Experimental observation of coherence resonance in an excitable chemical reaction system. *Physical Review E, 66*, 046204.

29. Beato, V., Sendina-Nadal, I., Gerdes, I., & Engel, H. (2008). Coherence resonance in a chemical excitable system driven by coloured noise. *Philosophical Transactions A, 366*, 381–395.

30. Gu, H., Yang, M., Li, L., Liu, Z., & Ren, W. (2002). Experimental observation of the stochastic bursting caused by coherence resonance in a neural pacemaker. *NeuroReport, 13*, 1657–1660.

31. Bertotti, G. & Mayergoyz, I. (2005). *The science of hysteresis*. Oxford: Elsevier.

32. Dimian, M., & Andrei, P. (2011). Noise induced resonance phenomena in stochastically driven hysteretic systems. *Journal of Applied Physics, 109*, 07D330.

33. Dimian, M., Andrei, P., Manu, O., & Popa, V. (2011). Comparison of noise-induced resonances for different models of hysteresis. *IEEE Transactions on Magnetics, 47*(10), 3825–3828.

34. Dimian, M., Manu, O., Andrei, P. (2012). Influence of noise color on stochastic resonance in hysteretic systems. *Journal of Applied Physics, 111*, 07D132.

35. Mantegna, R. N., Spagnolo, B., Testa, L., Trapanese, M. (2005). Stochastic resonance in magnetic systems described by Preisach hysteresis model. *Journal of Applied Physics, 97*, 10E519.

36. Testa, L., & Trapanese, M. (2008). Magnetic stochastic resonance in systems described by dynamic Preisach model. *Physica B, 403*, 486–490.

37. Gardiner, C. (2009). *Stochastic methods-a handbook for the natural and social sciences*. Berlin: Springer.

38. Melnikov, V. I. (1993). Schmitt trigger: A solvable model of stochastic resonance. *Physical Review E, 48*, 2481–2489.

39. Korman, C., & Mayergoyz, I. (1997). Review of Preisach type models driven by stochastic inputs as models for after-effect. *Physica B, 233*, 381–389.

40. Makra, P., Gingl, Z., & Fulei, T. (2003). Signal-to-noise ratio gain in stochastic resonators driven by coloured noises. *Physical Letters A, 317*, 228–232.

41. Fuentes, M. A., & Wio, H. S. (2006). Stochastic Resonance: influence of f^{-k} noise spectrum. *European Physical Journal B, 52*(2), 249–253.

Appendix A
Hysteresis Modeling in HysterSoft©

HysterSoft© is a computer program for the simulation of hysteresis and related phenomena in hysteretic systems (http://www.eng.fsu.edu/ms/HysterSoft). It provides a user-friendly simulation framework, in which various mathematical models of hysteresis can be implemented numerically relatively easy. HysterSoft© version 1.0 comes with the following models already implemented by default (these models are called *predefined models*): the backlash operator, the Bouc-Wen model, the Coleman-Hodgdon model, the stop (elastic-plastic) operator, the energetic model, the Jiles-Atherton model, the Langevin model, the limiting-loop proximity model, the Preisach model (including the cases in which the Preisach distribution is specified analytically or using discrete values defined on a mesh), and the rectangular-loop model. Users can also add additional models to the program by using predefined templates. These models are called *user-defined models*.

HysterSoft© can also be used to compute first-order reversal-curves (FORCs) diagrams, identify the model parameters from experimental data, conduct temperature and stress dependent simulations, perform noise passage analysis in hysteretic systems, etc. Most of these simulations can be performed by any hysteretic model either predefined or user-defined. The program can also be used as a dll library and called from other programs such as Matlab, Simulink, or C++.

In addition, HysterSoft© can be used to define hysteresis models for the electric permittivity and magnetic permeability. These models can be imported directly in RandFlux© and used to simulate electronic circuits containing hysteretic inductors, transformers, and capacitors.

A list of the most important features in HysterSoft© is given below:

1. Direct and inverse hysteresis modeling
2. Dynamic hysteresis modeling (e.g. frequency-dependent hysteresis modeling)
3. FORCs computations
4. Identify the Preisach distribution function using an experimental set of FORCs in the case of the scalar Preisach model and a set of FORCs measured at different angles in the case of the vector Preisach model.
5. Vector hysteresis modeling

M. Dimian and P. Andrei, *Noise-Driven Phenomena in Hysteretic Systems*,
Signals and Communication Technology 218, DOI: 10.1007/978-1-4614-1374-5,
© Springer Science+Business Media New York 2014

6. Noise passage analysis in hysteretic systems
7. Stochastic analysis of hysteretic systems
8. Thermal relaxation in both scalar and vector models
9. User-defined models of hysteresis
10. Possibility to use HysterSoft© as a library
11. Parameters identification tools for all the models (including the user-defined models)
12. Perform "loop simulations", which are more complex simulations defined by the user.

A.1 Simulations Using Scalar Models of Hysteresis

HysterSoft© can perform different types of simulations by using the modules implemented in the program. These types of simulations can be performed using any hysteresis model, predefined or user-defined. The modules can be selected from the main combo-box in the scalar model window.

A.1.1 Modules Available for Scalar Models of Hysteresis

Module A.   (*Input and output are defined using the GUI*) Simulate hysteresis curves be varying the input in the case of direct hysteresis modeling or output in the case of inverse hysteresis modeling using the track bars from the graphical user interface. This module can be used to simulate static hysteresis curves, visualize the current hysteretic state (e.g. the Preisach plane) in real time, compute the major hysteresis loop, initial and anhysteretic curves.

Module B.   (*Input defined from file*) Simulate hysteresis curves using the values of the input in the case of direct hysteresis modeling or output in the case of inverse hysteresis modeling defined in a file or in a textbox. The input or output should be defined one value per line. HysterSoft© will parse these values and compute the hysteretic curves. The user can use symbolic notations (or user-defined parameters) for the main physical parameters like in the following listing:

```
//sample input file
0
P1
−Hc
100
```

In the above example HysterSoft© will start from 0, apply an input equal to the value of user-defined parameter P1 (which is set in the options property grid), then another value equal to minus the coercive field of the material, and finally 100. The final values of the input and output are stored in the `FinalX` and `FinalY` variables. This module can be used in loop simulations, but cannot be used to perform dynamic simulations.

Module C.   (*Input defined analytically*) Compute the output variable if the input is a function of time defined analytically. The input signal should be defined using the `InputSignal` editor-. When defining the input signal it is convenient to use variables such as `tMin`, `tMax`, `Hc`, `P1`, `P2`, `P3`, `P4`, etc. in order to easily modify the signal from the property grid. It is also convenient to use such variables when performing loop simulations.

This module allows to perform rate-dependent simulations by setting the `EffectiveField` or `RelaxationTime` variables. The relaxation time should be a real number, while the effective field should be a string variable that defines the effective field as a function of the normalized rate (i.e. time derivative) of the output variable. For instance, by setting the `EffectiveField` variable to

`1.2*dm`

HysterSoft© will assume that the input of the system is equal to

$$x_{eff}(t) = x(t) + 1.2\frac{\dot{y}(t)}{y_s} \tag{0.1}$$

Rate-dependent simulations are particularly convenient for simulating frequency-dependent hysteresis.

Module D.   (*FORCs computation*) Compute a set of first-order reversal-curves. The user can define the maximum applied input (`XMax`), the number of reversal curves (`Reversals`), and the number of points per curve (`Resolution`) in the property grid. Dynamic models cannot be used when computing FORCs. If one needs to compute FORCs using a dynamic model it is recommended to define the input signal analytically by using the previous module (Module C).

Module E.   (*Thermal relaxation*) Perform simulations involving thermal relaxation. Thermal relaxation simulations are performed by adding a "thermal noise" to the input of a hysteresis model. The initial state can be obtained in two ways. The first way is to bring the hysteretic system in a given state by running a simulation with any other module (for instance modules A–C), then copying the hysteretic state, and using it in the thermal relaxation module. The second way to set the initial state is to use the `FirstFieldToApply` and `SecondFieldToApply` variables: before any thermal relaxation simulation HysterSoft© starts from the zero anhysteretic state and applies an input equal to `FirstFieldToApply`, then another input equal to `SecondFieldToApply`, and, then, performs the thermal relaxation simulations. If one does not want to use these variables (and, for instance, use the current hysteretic state in the memory) one needs to set the `FirstFieldToApply` and `SecondFieldToApply` to NaN.

The user can select different types of noise to use in the thermal relaxation including Ornstein-Uhlenbeck, Gaussian, Laplace, Cauchy, and uniformly distributed noise by setting the `TypeOfNoise` variable. For each type of noise the user can set the magnitude of the noise and different other parameters characteristic to the noise selected. In addition, the user can change the spectrum of the noise by changing the `NormalizeInputSpectrum` variable. By default the `NormalizeInputSpectrum` is set to 1, which means that the input noise is multiplied by 1 (i.e. left unchanged), however the user can change this variable to a frequency dependent function in order to modify the color of the noise. In this way, one can use colored noise inputs such as pink, blue, or violet noises.

The user can define the number of total averages for which the expected values of the output variable is computed, the experiment time, and the time step of the thermal noise. After the simulation is performed HysterSoft© will set the `ViscosityCoefficient` variable to the value of the viscosity coefficient.

This module can be used in loop simulations.

Module F.   (*Noise passage analysis*) Analyze the noise passage characteristics through hysteretic systems. This module allows the user to compute the output spectrum of any model of hysteresis if the input spectrum is given. The input spectrum can be of any type described in the previous module. This module can be used in loop simulations.

Module G.   (*Stochastic resonance*) Perform stochastic resonance simulations. This module can be used to compute the signal amplification (SA) and signal-to-noise ratio (SNR) in stochastic resonance. By default, the input signal is sinusoidal, but can be changed using the Signal Editor. In addition, parameters P1 and P2 can be used to easily modify the input signal, for instance when performing loop simulations. The input noise is defined in the same way like in the previous two modules. The signal amplification and signal-to-noise ratio are stored in the SA and SNR variables and can be used in loop simulations.

A.1.2 Loop Simulations

Loop simulations refer to simulations that are performed multiple times, each time for a different set of model parameters or input signals. For instance, such simulations are convenient to perform when computing the anhysteretic curve, when computing hysteresis curves for different values of model parameters, or when simulating the thermal relaxation, noise passage, or stochastic resonance for different parameters of the input noise.

Loop simulation can be performed using the "Loop simulation" editor. By default, HysterSoft© performs two embraced iteration loops for two different parameters. The parameters and the values that these parameters take during each simulation can be set in the "Major loop" and "Minor loop" tabs of the "Loop simulation" editor. The user can specify what variables to save on the hard disk, what curves from the output window to save, and what image files to produce after each iteration. The file names can be defined using parameters between two $ characters. For instance

```
Output_Hc_$Hc$.dat
```

will be expended by replacing Hc with the current value of the coercive field. Most of the figures showing simulation results in this book are produced using the loop simulation module in HysterSoft©.

A.1.3 Inverse Modeling

Inverse modeling can be performed by using the first two modules (Module A and Module B). Using inverse modeling the user needs to specify the values of the output for which the input needs to be computed and set the ModelingType variable to InverseModeling.

In the case of user-defined models, inverse modeling can be performed only if the model implements the Susceptibility or the ChangeY functions (see Sect. A.3).

A.1.4 Parameter Identification Tools

HysterSoft© implements three types of methods to determine the model parameters:

1. evolutionary identification techniques such as swarm optimization, genetic algorithms, and the Nelder-Mead method;
2. the Levenberg–Marquardt algorithm;
3. iterative techniques.

In addition, HysterSoft© implements model specific, parameter identification tools (for instance in the case of the energetic and Jiles-Atherton models). The values of the physical parameters that should be used in the identification problem (such as the coercive field, the output saturation, the remanence, initial susceptibility, etc.) can be specified in the graphical user interface.

In the case of user-defined models, the model parameters that appear in the parameter identification editor should be identified using the [IdentificationModelParameter] attribute (see Sect. A.3).

A.2   Simulations Using Vector Models of Hysteresis

By default, HysterSoft© adds a two-dimensional vector model of hysteresis to any scalar model. The vector model is defined using superposition as explained in Chap. 1. Similar to the case of scalar models of hysteresis, HysterSoft© has a number of modules that can be used to perform vector simulations. These modules are summarized below.

Module A.   (*Input and output are defined using the GUI*) Simulate vector hysteresis curves be varying the input using the buttons on the graphical user interface (these buttons can also be pressed using the arrow buttons on the keyboard).

Module B.   (*Input defined from file*) Simulate hysteresis curves be using the input from a file or from a textbox. Both the x and y components of the input should be defined one the same line. HysterSoft© will parse these values and compute the output variable. Shortcuts for various parameters can also be used like in the next example

```
//sample input file for vector simulations
0, 0
HcX, 0
HcX, HcY
P1, P2
0, -P3
```

In this example HysterSoft© starts from $(x_x, x_y) = (0, 0)$, than applies (HcX, 0), (HcX, HcY), (P1, P2), and finally (0, −P3). The values of HcX, HxY, P1, P2, P3, etc. will be expended to the numerical values defined by the user in the property grid. The final value of the input and output components after the simulation are given by the FinalX_x, FinalX_y, FinalY_x, and FinalY_x variables. These variables can be used in loop simulations.

Module C.   (*FORCs computation*) Compute vectorial FORCs, by specifying the maximum applied input (XMax), the number of reversal curves (Reversals), the number of points per curve (Resolution), and the angle (Angle) under which the FORCs are computed.

Module D.

These values can all be specified in the property grid. Dynamic models cannot be used when computing vectorial FORCs.

(*Thermal relaxation*) Perform vectorial thermal relaxation simulations. The initial hysteretic state can be specified by using the copy and paste features like in the case of scalar models. The user can also specify the magnitude of the noise, the number of averages, and the number of points in which the thermal relaxation is computed.

The parameters of the scalar hysteresis model in any particular direction are computed using the values of the parameters in the x and y directions, as explained in Sect. 1.8.1. The parameters in the x and y directions can be defined by the user using the model property grid.

A.3   Defining New Hysteresis Models

New scalar and vector hysteresis models can be easily defined in HysterSoft©. To define a new scalar model, start the "New Scalar Model" editor from the "Scalar Models" menu. Using this editor one can define the equations of the new model, properties that should appear in the property grid, default values for the model parameters, the parameters that appear in the parameter identification tool, etc. The easiest way to learn how to define new hysteresis models is to start with the examples that come with the installation kit.

The code of the new model should be written in C# and should contain the definition of the model as a C# class. This class should be inherited from the `UserDefinedModel` class (which is internally implemented in HysterSoft©). Any user-defined model should define at least one of the following two functions

```
double Susceptibility(double x, double y, int
sensOfVariation)
```
or
```
double ChangeX(double x)
```

The first function defines the susceptibility of the model, while the second describes how the output variable changes when the input becomes equal to x). The first function is convenient for use in differential models of hysteresis, while the second one in algebraic models. Once any of the above functions is implemented the user can perform any of the types of simulations presented above, including thermal relaxation and noise passage analysis.

Now let us look at an example that comes by default in HysterSoft©. This example implements the Jiles-Atherton model.

```csharp
public class MyModel : UserDefinedModel
{
    //a default constructor is optional
    public MyModel()
    {
        xMax = 100;
    }

    //Define the differential susceptibility
    public override double Susceptibility(double x,
double y, int sensOfVariation)
    {
        double l, dl;
        l = SpecialFunctions.Langevin(x+alpha*y)-y;
        dl = SpecialFunctions.dLangevin(x+alpha*y);

        if (sensOfVariation*l <= 0)
            return c*dl;

        //return the differential susceptibility
        return (1-c)*l/(k*(1-c)*sensOfVariation-
alpha*dl)+c*dl;
    }

    //Add some properties on the property grid
    private double a = 20;
    [IdentificationModelParameter(1, 100)]
    public double _a { get{return a;} set{a=value;} }

    private double c = 0.1;
    [IdentificationModelParameter(0, 1)]
    public double _c { get{return c;} set{c=value;} }

    private double k = 40;
    [IdentificationModelParameter(1, 100)]
    public double _k { get{return k;} set{k=value;} }

    private double alpha = 0;
    [IdentificationModelParameter(0, 1e-5)]
    public double _alpha { get{return alpha;}
set{alpha=value;} }

}
```

The Susceptibility function is mandatory (because we do not implement the ChangeX function). The rest of all other functions and properties are optional, including the default constructor.

The [IdentificationModelParameter(...)] attribute for properties is also optional and tells HysterSoft© that the given property defines a model parameter that should appear in the Parameter Identification Tool (see Sect. A.1.4). The two parameters of the IdentificationModelParameter attribute denote the minimum and maximum values within which HysterSoft© can search during the parameter identification.

Functions SpecialFunctions.Langevin and SpecialFunctions. dLangevin are internally defined in HysterSoft© and they compute the Langevin function and the derivative of the Langevin function. The user can use his or her own implementations for these functions.

The ToString function defines a name of the current model.

Next let us look at another example that implements the Langevin model, which is an algebraic model.

```
public class MyModel : UserDefinedModel
{
    public override double ChangeX(double x)
    {
        return Ms*SpecialFunctions.Langevin(x/3);
    }
}
```

Notice that HysterSoft© will always define a variable Ms, which is by default equal to 1. Since hysteresis is a history dependent phenomenon, HysterSoft© allows users to recall the previous values of the input and output variables. The following predefined variables can be used from anywhere inside the class definition (including from inside functions Susceptibility and ChangeX):

- state.X – the last value of the input
- state.Y – the last value of the output
- state.XReversal – the last value of the input reversal
- state.YReversal – the last value of the input reversal
- state.pastX[i] – the previous i-th value of the input
- state.pastY[i] – the previous i-th value of the output

Notice in the last example that HysterSoft© will not be able to perform inverse modeling simulations because the class does not implement the Susceptibility or the ChangeY functions.

A.4   Saving the Model Parameters

The parameters of any model of hysteresis can be saved in xml files that later can be re-loaded and used in other simulations. By default, HysterSoft© assigns the .hyst file extension to any model parameter file. Models parameters files are in general compatible from one version of HysterSoft© to another.

A.5   Computing the Scalar Preisach Distribution Function From First-Order Reversal-Curves

The "Scalar FORCs Analysis" tool in HysterSoft© can be used to compute the Everett distribution and the reversible $R(\alpha)$ and irreversible $P(\alpha, \beta)$ Preisach distributions if a set of FORCs is provided. The set of FORCs should be given as a 3-column file giving the values of the reversal field, current field, and current output for each curve, like in the example below:

```
10  10  1        : this is the first curve which consists of one point
9   9   0.99     : here starts the second reversal-curve which consists of 2 points
9   10  1
8   8   0.98     : here starts the third reversal-curve which consists of 3 points
8   9   0.99
8   10  1
. . .
-10  -10  -1     : here starts the last reversal-curve
-10  -9   -1
. . .
-10  10   1
```

The user can specify any number of reversal curves, each having any number of points. The reversal fields and the points on each reversal curve can be distributed non-uniformly, however, in order to increase the accuracy of computations it is recommended to use uniform distributions. HysterSoft© will automatically detect the number of reversal curves and points on each curve. FORCs should be given as ASCII files with extension .forcs.

The Everett distribution can be saved as a .everett file, while the computed Preisach distribution as a .preisach file. The structure of the .preisach file is similar to the structure of the .forcs file and consists of 3 columns giving the values of the reversible component of the Preisach distribution $R(\alpha)$ when $\alpha = \beta$ and the irreversible component of the Preisach distribution irreversible $P(\alpha, \beta)$ when $\alpha > \beta$. This is a sample .preisach file

```
10  10  0.01    : R(10) = 0.01
9   9   0.02    : R(9) = 0.02
9   10  0       : the first point of the irreversible component P(9,10) = 0
8   8   0.03    : R(8) = 0.03
8   9   0.1     : the next point of the irreversible component P(8,9) = 0.1
8   10  0.02    : the next point of the irreversible component P(8,10) = 0.02
. . .
-10 -10 0.1  : R(-10) = 0.1

-10 -9  -0.1 : P(-10,-9) = 0.1

. . .
-10 10  0.1     : P(-10,10) = 0.1
```

The .everett and .forcs files can be loaded by the Preisach model with discrete Preisach distributions and used to perform frequency-dependent simulations or thermal and noise analysis.

To identify the Preisach distributions of the vector Preisach model, HysterSoft© allows users to load multiple .forcs files and vector distributions. This can be done using the "Load FORCs Analysis" and "Load Multiple FORCs" tools.

A.6   Computing the Vector Preisach Distribution Function from a Set of First-Order Reversal-Curves

To identify the Preisach distributions of the vector Preisach model, HysterSoft© allows users to load multiple .forcs files and vector distributions. This can be done using the "Load FORCs Analysis" and "Load Multiple FORCs" tools. The FORCs in each angular direction need to be specified in different files and loaded in the software. HysterSoft© can compute the Preisach distribution function for both two-dimensional and three-dimensional FORCs in the case of isotropic hysteretic systems, and only for two-dimensional FORCs in the case of anisotropic systems. Notice that two-dimensional FORCs should be specified along each polar angle φ, while three-dimensional FORCs should be specified along each polar and azimuthal angles φ and θ

A.7   Performing Temperature and Stress Dependent Simulations

Temperature and mechanical stress change the shape of the hysteresis loops in most magnetic and electric hysteretic systems. For instance, in the case of magnetic systems, increasing the temperature will usually decrease the value of the coercive field, output remanence and saturation, while increasing the stress will increase the output remanence and make the shape of the major loop more rectangular. Temperature and stress dependent hysteresis curves can be computed using the energetic and Jiles-Atherton models of hysteresis. The values of the temperature and stress can be specified in the parameter window of the model.

A.8   Using HysterSoft© as a Library

Most HysterSoft© functions can be called from other Windows applications (such as Matlab, C, Fortran, etc.). There are two ways to call HysterSoft© functions from other programs:

- load the HysteresisLibrary.dll file as a .NET assembly, selecting the desired hysteresis model, and call the public functions and properties (recommended)
- use HysterSoft as a COM object (not recommended; this way is obsolete and is kept for compatibility with old versions of HysterSoft©).

Below there are a few examples that show how to call HysterSoft© from Matlab by loading HysteresisLibrary.dll as an assembly. These examples can also be found in the /Matlab subdirectory from the installation directory of HysterSoft©.

Example 1: Loading the HysterSoft library, selecting a model, setting model parameters, and computing the values of some physical parameters.

```
%load the HysterSoft library
a = NET.addAssembly('HysteresisLibrary.dll')

%use the Jiles-Atherton model
ja = HysterSoft.JilesModel()

%reset simulations (optional)
ja.Reset()

%set some parameters (optional)
ja.a_ = 40
ja.alpha_ = 0
ja.c_ = 0.15
ja.k_ = 30
ja.Ms_ = 1

%compute the values of physical parameters
ja.Hc        %coercive field
ja.Mr        %remanent output
ja.SusIni    %initial susceptibility
ja.SusC      %susceptibility at coercivity
ja.SusR      %susceptibility at remanence

%perform some computations
ja.ChangeH(123)
ja.ChangeH(-10)
```

Note that the model parameters have an underscore character after the parameter name. This is because variables in Matlab cannot start with the underscore character, so all the variables that start with an underscore in HysterSoft© will be renamed in Matlab.

The following table shows the class names of the scalar hysteresis models currently implemented in HysterSoft©.

Model	Command
Backlash operator	`BacklashOperator()`
Bouc-Wen	`BoucWenModel()`
Coleman-Hodgdon	`HodgdonModel()`
Elastic-plastic operator	`ElasticPlasticOperator()`
Energetic (Hauser)	`EnergeticModel()`
Jiles-Atherton	`JilesModel()`
Langevin	`LangevinModel()`
Limiting-loop proximity	`LimitingLoopProximity()`
Preisach	`JilesModel()`
Rectangular loop	`RectangularLoopModel()`

Model	Command
Backlash operator	`VectorBacklashOperator()`
Bouc-Wen	`VectorBoucWenModel()`
Coleman-Hodgdon	`VectorHodgdonModel()`
Elastic-plastic operator	`VectorElasticPlasticOperator()`
Energetic (Hauser)	`VectorEnergeticModel()`
Jiles-Atherton	`VectorJilesModel()`
Langevin	`VectorLangevinModel()`
Limiting-loop proximity	`VectorLimitingLoopProximity()`
Preisach	`VectorPreisachModel()`
Rectangular loop	`VectorRectangularLoopModel()`

Example 2: Solving inverse problems.

```
a = NET.addAssembly('HysteresisLibrary.dll')
ja = HysterSoft.JilesModel()

ja.ChangeM(-0.5)
```

Example 3: Ploting the major hysteresis loop.

```
a = NET.addAssembly('HysteresisLibrary.dll')
ja = HysterSoft.JilesModel()

x  = 300:-1:-300
yU = arrayfun(@ja.ChangeH, x)
yL = arrayfun(@ja.ChangeH, -x)
plot(x,yL, -x,yU)
```

Example 4: Saving and loading material parameters, displaying the hysteretic state window, and the parameter identification window.

```
a =  NET.addAssembly('HysteresisLibrary.dll')
ja = HysterSoft.JilesModel()

%save the model parameters in a .xml file
ja.SaveMaterial('fileName.xml')

%load the model parameters from a .xml file
ja.LoadMaterial('fileName.xml')

%show the hysteretic  state window
ja.lsp.ShowState()

%call the parameter identificatiion tool
ja.IterativeIdentificationTool()
```

Example 5: Working with vector models. The following example shows how to use the 2-D vector Preisach model.

The following table shows the class names of the vector hysteresis models currently implemented in HysterSoft©.

```
%load the HysterSoft library
a =  NET.addAssembly('HysteresisLibrary.dll')

%use the Jiles-Atherton model
pm = HysterSoft.VectorPreisachModel()

%reset simulations (optional)
pm.Reset()

%set the type of the distribution and the model parame-
ters along the x-axis and y-axis (optional)
pm.TypeInteractions = HysterSoft.InterDisType.Normal
pm.TypeCoercivities = HysterSoft.CoercDisType.LogNormal
pm.TypeReversible = HysterSoft.RevDisType.Cauchy

pm.HC01_ = 100
pm.HSC1_ = 30
pm.HSI1_ = 30
pm.HSRev1_ = 70
pm.S1_ = 0.9

pm.HC02_ = 50
pm.HSC2_ = 20
pm.HSI2_ = 30
pm.HSRev2_ = 70
pm.S2_ = 0.9


%change the input variable and display the output
y = pm.ChangeH(10,10)

y(1) %this is the x-component of the output
y(2) %this is the y-component of the output

%plot a circular loop
t = [0:0.1:2*pi];
input_x = 50*sin(t);
input_y = 50*cos(t);
output_x = zeros(1,length(t));
output_y = zeros(1,length(t));
for i = 1:length(t)
    y = pm.ChangeH(input_x(i),input_y(i));
    output_x(i) = y(1);
    output_y(i) = y(2);
end
plot(output_x, output_y)
```

Printed in the United States
By Bookmasters